COMPLEX STOCHASTIC SYSTEMS

Printed in Great Britain for the Royal Society
by the
University Press, Cambridge

Book: ISBN 0 85403 453 6
Journal: ISSN 0080 4614

Also published in *Philosophical Transactions of the Royal Society of London*, series A, volume 337 (no. 1647), pages 307–428

The text paper used in this publication is alkaline sized with a coating which is predominantly calcium carbonate. The resultant surface pH is in excess of 7.5, which gives maximum practical permanence.

British Library Cataloguing-in-Publication Data

A CIP catalogue record for this
book is available from
the British Library

ISBN 0-85403-453-6

Published by the Royal Society
6 Carlton House Terrace, London SW1Y 5AG

Complex stochastic systems

Compiled and edited by D. R. Cox and D. M. Titterington

PAGE

Phil. Trans. R. Soc. Lond. A (1991) **337**, 307–428
Printed in Great Britain

13-2

Introductory remarks

By D. R. Cox[1] and D. M. Titterington[2]

[1] *Nuffield College, Oxford OX1 1NF, U.K.*

[2] *Department of Statistics, University of Glasgow, University Gardens, Glasgow G12 8QQ, U.K.*

Current research on statistical topics spans a wide range of ideas and fields of application. Much is connected with the systematic theory of methods for the analysis of empirical data, especially for situations, common in many areas of science and technology, where the random or haphazard element in the data is too strong to be ignored. Modern computer technology is important in enabling large amounts of data to be explored quickly and efficiently, in allowing methods involving iterative calculations to be handled as a matter of routine and in facilitating display of the results of analysis via sophisticated graphical devices. This has allowed both the development of new methods and also the application of ideas long known in principle but until relatively recently too complicated for other than occasional use. Some of the key ideas involved go back in essence to the 19th century, for example to Gauss, and others are more strongly associated with the first half of the 20th century, in particular to the pioneering work of R. A. Fisher, although many of the detailed developments are, of course, much more recent.

In a further band of applications the complex structure and large extent of the data are of the essence rather than an incidental complication. In 1986 the Science and Engineering Research Council recognized the existence of these newer kinds of application by launching an Initiative under the name Complex Stochastic Systems.

The papers that follow represent a fairly small part of the work supported under that Initiative.

F. P. Kelly develops some of the elegant probability theory required for studying the properties and control of large networks. The immediate application is to telecommunication systems but the implications are broader, both in an engineering and an economic context, for example. Image analysis yields many important instances of complex stochastic systems. Major progress has of course been possible by relatively informal methods. The challenge to a more formal statistical approach is to provide methods both based on a stochastic model for the true 'scene' under study and at the same time computationally feasible. The papers of W. Qian & D. M. Titterington and R. J. Aykroyd & P. J. Green provide examples of this line of work and illustrate some of the areas of application that benefit from the new methods.

A major source of the 'complexity' of the above topics is the need to reflect lack of independence in some relevant sense. This feature is also present in the study of expert systems and the paper by D. J. Spiegelhalter, A. P. Dawid, T. A. Hutchinson & R. G. Cowell provides and illustrates a probabilistic approach to this important topic.

Modern instrumentation frequently yields automatically large amounts of multidimensional data and the calibration and reduction of such data raises special

Phil. Trans. R. Soc. Lond. A (1991) **337**, 309–310

Printed in Great Britain

problems, some aspects of which are addressed by P. J. Brown, C. H. Spiegelman and M. C. Denham.

Many complex stochastic systems require for their numerical study the evaluation of high-dimensional multiple integrals and techniques for this are discussed by A. F. M. Smith.

Other kinds of complex stochastic system not represented in the present issue include neural nets, (chaotic) nonlinear time series, aspects of signal processing and the mathematical modelling of population genetics.

Chemometrics and spectral frequency selection

By Philip J. Brown[1], Clifford H. Spiegelman[2] and Michael C. Denham[1]

[1] *Department of Statistics and Computational Mathematics, University of Liverpool, PO Box 147, Liverpool L69 3BX, U.K.*

[2] *Department of Statistics, Texas A & M, College Station, Texas 77843, U.S.A.*

In many fields of science, the simple straight line has received more attention as a basis for calibration than any other form. This is because measuring devices have been mainly univariate and have had calibration curves which were sufficiently linear. As scientific fields become more computationally intensive they rely on more computer-driven multivariate measurement devices. The number of responses may be large. For example modern scanning infrared (IR) spectroscopes measure the absorptions or reflectances at a sequence of around one thousand frequencies. Training data may consist of the order of 10 to 100 carefully designed samples for which the true composition is either known by formulation or accurately determined by wet chemistry. In future one wishes to predict the true composition from the spectrum. In this paper we develop a variable selection approach which is both simple in concept and computationally easy to implement. Its motivation is the minimization of the width of a confidence interval. The technique for data reduction is illustrated on a mid-IR spectroscopic analysis of a liquid detergent in which the calibrating data consists of 12 observations of absorptions at 1168 frequency channels (responses) corresponding to five chemical ingredients.

1. Introduction

1.1. *Spectroscopic data*

Infrared spectroscopy involves directing a pulse of infrared (IR) light onto a substance, typically but not exclusively a liquid, noting the energizing effect over a short interval of time, and converting this to absorbances at various frequencies. Modern scanning instruments allow the simultaneous examination of a frequency range of around 1000 contiguous frequencies. The absorbances plotted at these frequencies represent the spectrum for that liquid sample. The 12 mid-IR spectra samples plotted over 1168 distinct frequencies are presented in figure 1. The samples are different mixtures of four detergent ingredients in aqueous solution. Each mixture involves a different amount of each of the five constituents. These plots appear continuous up to the resolution of the plotting device but are in reality 1168 discrete points. The 12 plots appear very similar at first sight. These data make up our calibrating data used to fit and analyse the models of this paper.

Figure 2 gives the sample mean and variance of the mid-IR spectra for the calibration data, averaging over the 12 observations. The 12 curves are broadly of a similar form to the mean curve, but there is also considerable variation, usually at those frequencies where absorbances are high. In fact we shall see that very accurate predictions of concentrations of ingredients may be obtained from these curves.

Phil. Trans. R. Soc. Lond. A (1991) **337**, 311–322

Printed in Great Britain

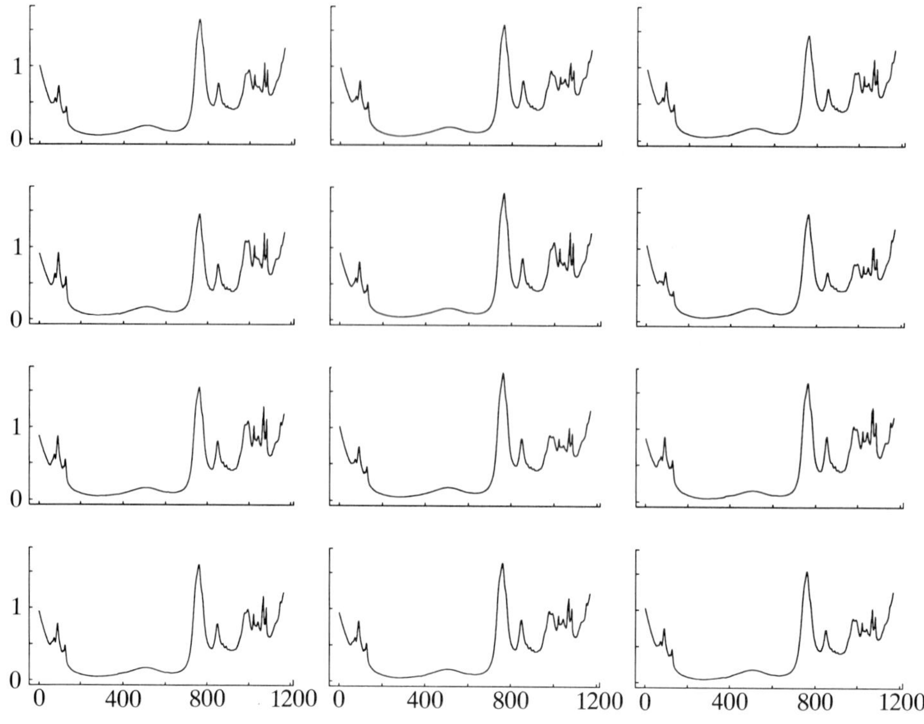

Figure 1. Absorbance spectra for the 12 detergent samples.

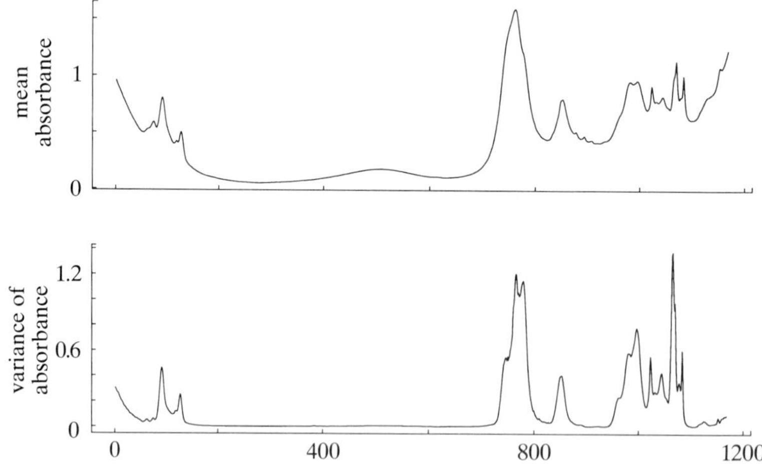

Figure 2. Mean and variance of 12 absorbance spectra.

The relationship of spectrum to ingredients may be motivated as follows. Briefly, molecules of a substance are able to absorb the incident IR radiation by moving between different vibrational and rotational energy levels of the lowest electronic energy state. In the very simple case of a diatomic A–B molecule the only vibration that can occur is a periodic stretching along the A–B bond. These stretching motions allow the vibrational frequency to be approximately predicted from Hooke's law. It is possible to consider the vibrations of individual bonds in more complex molecules in a similar manner, but other forms of vibration of individual bonds become possible, including rocking, scissoring, twisting and wagging (see, for example, Cross & Jones 1969). Spectroscopists have verified that specific absorption bands for particular bonds or groups within a molecule occur at, or near, the expected frequencies. All these vibrational mode absorption frequencies tend to be altered in varying degrees by small changes in the remainder of the molecule, leading to highly characteristic IR spectra for organic substances.

The IR absorption of a particular mixture will be related to what chemicals are present and in what amounts. In fact in many circumstances the relationship has been found to be linear. This relationship is characterized by Beer's law (Cross & Jones 1969), which stipulates a linear relationship between the concentration of a substance dispersed in a non-absorbing medium and the amount of light absorbed by it. This is only an experimental law and departures from linearity occur for a variety of reasons, including lack of dispersion throughout the medium and scattering effects in the case of particulate solids.

Although the spectroscopist can identify bands of frequencies which give high absorbances for particular ingredients it is much harder to specify *a priori* those frequencies which best discriminate between the ingredients. Modern methods of chemometric analysis usually avoid such selection or apply some automated method. It is the purpose of this paper to explore a new method of frequency selection which is both easy to apply and demonstrated to be effective. This method is also able to cope with the high degree of indeterminacy in typical spectroscopic data where there are many more variables than observations.

1.2. *Regression models and overfitting*

There are two modes of approaching the fitting of a relationship between the spectrum ($Y, 1168 \times 1$) and the detergent and water concentrations ($x, 5 \times 1$). Suppose we regress Y on x. To predict x in future we may invert the fitted relationship, taking account of covariance structure. Or, more directly, we may regress x on Y so as to be able to predict x from Y in future. The latter is often favoured for its simplicity in the chemometrics literature. However, when Y is the response and embodies the error, statisticians have long advocated regressing in the direction of the error (see Eisenhart 1939; Williams 1969; Brown 1982). When there are fewer observations (12) than spectral variables (1168) then both approaches lead to the same indeterminate fit, with the same non-unique solution subspace (see Sundberg & Brown 1989). However, despite the conforming degeneracy, the focus of our approach is largely that of regression of spectra on concentrations, as we regard this as conceptually most satisfactory when the concentration data are designed in calibration as in the present example.

In such an indeterminate, over-parametrized problem there are a variety of ways to regularize and achieve stable estimates: partial least squares regression (see Martens & Naes 1989; Brown 1990); principal components regression; ridge regression (Marquardt 1979); coherent calibration (Brown & Mäkeläinen 1992); minimum length least squares (using the Moore–Penrose generalized inverse); and continuum regression (Stone & Brooks 1990). None of these methods is unique. They depend critically on the scales of the variables used to predict. All assume implicitly, or explicitly in the case of Brown & Mäkeläinen (1992), some sort of prior assumptions for the true regression coefficients, and will fare more or less well depending on whether or not such implicit assumptions approximately hold, which can be influenced by the choice of transformation. Brown & Mäkeläinen (1992) and Denham

& Brown (1990) utilize the fact that absorbance is a continuous function of frequency, the former assuming a stationary gaussian process prior distribution, the latter using splines and autoregressive error structures.

Although it is appealing to recognize that inferentially one can in principle always do better by using more information, since one is then always free to ignore the information, in scientific research there is a tradition which emphasizes the fragility of inference with large numbers of variables and the inevitable associated degree of overfitting. Even in the closed bayesian inferential system, tractable prior distributions for large numbers of parameters have unforeseen and undesirable features (see Dawid 1988; Mäkeläinen & Brown 1988). On the other hand, in the sampling framework of statistical inference, a set of complementary problems arise. A large number of estimates of effects which are truly zero will throw up at random a proportion of large estimates of seemingly large effects; the problem of multiple comparison. Regularization or shrinkage offers one answer, but may still be open to some overfitting. Internal leave-one-out cross-validation, a valuable tool against self-deception (see Stone 1974), may flatter in its internal cross-validated mean squared errors, especially as it assumes future prediction data that is very like that sampled. In a further application of the technique of this paper, using a validation set whose range is outside that of the calibrating data to predict the concentrations of three sugars in solution, the cross-validation mean squared errors were found to be an order of magnitude too small (Brown 1991).

In the approach we adopt here we remove a large number of frequencies so as to reduce vulnerability to overfitting. This is achieved by choosing frequencies to minimize the length of a confidence interval. Before specifying our method in §4 we review approaches to variable selection in the next section. The model adopted is presented in §3 and the application described in §5. The reader wishing to skip the technical details of §4 may skip to the end of that section for a brief summary of the method before proceeding to the example.

2. Variable selection reviewed

The large number of responses and the relative paucity of observations in modern calibration make many of the standard variable selection techniques used in regression inapplicable or computationally prohibitive. Commonly used methods of variable selection are all possible regressions, best subset selection, backward elimination, forward selection and stepwise regression. For details of these approaches the reader is referred to Miller (1990). Clearly, since backward elimination relies on including all variables in the regression model and then eliminating variables, it cannot be used due to the non-uniqueness of the regression equation based on including all q variables. The use of the all possible regressions approach is also infeasible. Even if we restrict attention to those subsets which give unique estimators, the number of models which must be considered is vast. Best subset regression, which requires a fraction of the time required by all possible regressions, will still consider too many regressions for it to be practicable in many cases.

The use of forward selection methods is widespread (Fearn 1983), and some instruments have such an approach incorporated in their accompanying software (see Osborne *et al.* 1984). Fearn (1983) gives an example of calibrating protein content in ground wheat where forward selection is found wanting, partly because of high correlations between absorbances at different frequencies and low correlation with the component to be predicted. The use of stepwise regression also suffers in a similar way when applied to this example.

Selection techniques geared to prediction of the controlled variable via the fitted regression of Y on x have the same computational drawbacks. Brown (1982) uses a test of 'additional information' to select a subset of variables, and Spezzaferri (1985) uses a bayesian information theoretic approach.

Sometimes authors use a variable selection technique after taking principal components. This does not, however, meet our aim of selecting frequencies since all of the original set of frequency channels are still retained.

So far in this section we have assumed that we are concerned with predicting only one component. With $p > 1$ components one may consider the separate predictions and take the union of the frequencies selected for each prediction. Looking for a frequency set which simultaneously discriminates among the p components may be preferable if the aim is to choose the smallest subset. It would require a weighted sum of residual mean squares as an overall measure of goodness. Such an approach would add considerably to the complexity and entail a heavy computational overhead.

In this paper we propose an extremely simple technique for response selection. It does not claim to give the smallest possible subset but it does provide a predictively good subset which substantially reduces the number of channels.

3. The model for selection

As discussed we adopt the approach of regression of the absorption spectra on composition. We also regress on only one ingredient at a time. Although it might superficially seem desirable to regress on all five ingredients simultaneously, or at least on four of them since they total 100%, we want subsequently to predict the concentration of each of the four detergent chemicals using just the spectrum without knowledge of concentrations of the other three chemicals in the sample, see Brown (1982) for further discussion.

For the moment assume the simpler linear regression models,

$$Y_{kj} = \alpha_j + \beta_j x_k + \epsilon_{kj}, \tag{1}$$

where $j = 1, \ldots, q$ indexes the $q = 1168$ frequencies, and $k = 1, \ldots, n$ the $n = 12$ observations. Here x_k is the concentration of one of the ingredients, for the kth sample. In this model, errors $\{\epsilon_{kj}\}$ have zero mean and variance σ_j^2 and are assumed to be uncorrelated. The error variance at frequency j, σ_j^2, incorporates both measurement error and residual effects due to omitted constituents of the sample. The uncorrelatedness of the errors is perhaps an over-simplification since correlation is induced by these omitted constituents. In this case one could entertain the possibility of correlation of Y_{kj} across channels within an observation, but with the not insubstantial additional problem of a very large unknown covariance matrix; see Denham & Brown (1990) for some possible ways of structuring such problems. If such correlations are substantial it will be important to use them explicitly for prediction or to use them implicitly by a regularized method, such as partial least squares, that regresses x on Y.

As already discussed, the formulation is in terms of controlled calibration, where the explanatory variables are of future interest in prediction. Controlled calibration is most appropriate when these explanatory variables are initially fixed in some designed calibration experiment, as in the present example.

The n observations $\{Y_{kj}, x_k; j = 1, \ldots, q; k = 1, \ldots, n\}$ serve to estimate the unknown parameters, $\alpha_j, \beta_j, \sigma_j^2$ in model (1). With these q estimated calibrated relationships we may predict any further unknown sample compositions $\xi_t, t = 1, \ldots, T$, where we have designated unknown x by ξ. For this prediction problem the analogous model to (1) is

$$Z_{tj} = \alpha_j + \beta_j \xi_t + \epsilon_{tj}, \tag{2}$$

where ϵ_{tj} are uncorrelated errors with variance σ_j^2 ($j = 1, \ldots, q$; $t = 1, \ldots, T$).

We use the single-component method to tackle the multicomponent situation as in our illustration in §5 by successively applying the single composition results marginally to the p components and amalgamating the p sets of selected wavelengths. That is, we look to discriminate in turn between each ingredient and the other ingredients taken together.

One approach to choice of model dimension is to choose the size of model by calculating the mean squared error of prediction. This mean squared error is the sum of a variance and a bias squared term. Larger models have increased variance but decreased bias and an optimum compromise between the two is possible. A related but different approach which we adopt focuses on prediction intervals. The fitted model from (1) is not the true model. It is estimated once but may be used repeatedly to predict future values of ξ, each prediction incorporating the same bias of fitting. In addition to this source of error from the calibrating experiment there is the error generated by the postcalibration or prediction experiment. These two sources of error are affected by the size of model or number, q, of frequency channels adopted and determine the width of the confidence interval. We do not here look at the simultaneous satisfaction of all future use prediction intervals with an ascribed probability, rather we focus marginally on each single use interval. Again we are able to achieve an optimal compromise which minimizes the width of interval and gives a unique choice of selected channels.

When q is large, the numerous systematic errors will give rise to a proportion which is excessively large and spawns a related large literature of methods which seek to shrink estimates towards zero.

Quite distinct but equally problematic considerations arise within a bayesian framework. Suppose $\xi_1, \ldots, \xi_T$ are viewed as exchangeable *a priori*, that is, having a joint distribution which is invariant to permutations of the indices and consequently not in general independent; then $Z_1, \ldots, Z_t$ provide information on the form of this exchangeability, and posterior inference about ξ_t will depend on $Z_1, \ldots, Z_{t-1}$ as well as Z_t and the training data. Additionally, the multiple use of the calibration with its prior necessitates a careful assessment of the stability of inference to alterations of this prior distribution (see Berry 1988).

For the selection of frequency channels in this paper we work within the sampling theory inferential framework. Our basic idea is to choose that subset of q' of the q channels such that the approximated length of each single use prediction interval is minimized. Both q' and the corresponding subset of frequencies are chosen by the method.

4. The selection method

Since x and ξ are scalars, we seek a linear combination of the response to each instrument. We can then use the well-developed univariate methodology. Henceforth in the prediction model (2) we refer to the tth future Z and drop the subscript t to it. Let $Z = (Z_1, \ldots, Z_q)^T$ and $Z_\theta = \Sigma\,\theta_j Z_j$. Similarly let $\alpha_\theta = \Sigma\,\theta_j \alpha_j$ and $\beta_\theta = \Sigma\,\theta_j \beta_j$. We look for values of $\theta = (\theta_1, \ldots, \theta_q)^T$, with typically $q' < q$ non-zero components, which minimize the approximate length of certain confidence intervals. In models (1), (2) in addition to the second-order error assumptions the errors are taken to be normally distributed. Readers wishing to skip the following technical derivation may proceed to the last paragraph of this section.

For prescribed θ, the compound response Z_θ is normally distributed with, as mean, the calibration line, $\theta^T E(Z) = \alpha_\theta + \beta_\theta \xi$, and variance $\Sigma \theta_j^2 \sigma_j^2$. We proceed by means of the Cauchy–Schwarz inequality. First, with prescribed confidence level $1-\gamma$ and fixed θ with q' specific non-zero components,

$$|Z_\theta - \Sigma \theta_j(\alpha_j + \beta_j \xi)| = |\Sigma \theta_j \sigma_j \epsilon_j^*| \leqslant \surd[\Sigma \theta_j^2 \sigma_j^2] \surd[\chi^2_{1-\gamma}(q')] \tag{3}$$

since ϵ_j^* are now independent standard normal. Secondly and similarly, with probability $1-\delta$,

$$|\Sigma \theta_j(\hat{\alpha}_j + \hat{\beta}_j \xi) - \Sigma \theta_j(\alpha_j + \beta_j \xi)| \leqslant \surd[\Sigma \theta_j^2 \sigma_j^2 s^2(\xi)] \surd[\chi^2_{1-\delta}(q')], \tag{4}$$

where $s^2(\xi) = [1/n + \{(\xi - \bar{x})^2 / \Sigma (x_k - \bar{x})^2\}]$ and q' is the number of prescribed non-zero θ_j. These statements will also be true for θ chosen by the calibrating data in model (1) provided the randomness induced does not effect the actual channels selected. We give sufficient conditions for this later.

Let B be the event represented by (3) and A the event represented by (4). Then

$$\begin{aligned} P(B|A) &= P(A,B)/P(A) \\ &= [P(A) + P(B) - P(A \cup B)]/P(A) \geqslant 1 + [(1-\gamma) - 1]/(1-\gamma) \\ &= 1 - \gamma/(1-\delta) \\ &\doteq 1 - \gamma. \end{aligned}$$

Thus, for 'good' calibrating events prescribed by (4) which occur with probability $(1-\delta)$, we have future predictions prescribed by (3) with probability approximately $(1-\gamma)$, the approximation being better the smaller the value of γ and δ. Inequality (4) may be adapted to give a statement for all future ξ, and the χ^2 degrees of freedom increase to $2q'$. With inequality (3) this would form a basis for simultaneous repeated-use confidence intervals (see also Scheffé 1973; Carroll, *et al.* 1988). We, however, prefer the single-use statements.

We let $c_1 = \surd[\chi^2_{1-\gamma}(q')]$ and $c_2 = \surd[\chi^2_{1-\delta}(q')]$. Now, the triangle inequality gives $|Z_\theta - \hat{E}(Z_\theta)| \leqslant |Z_\theta - E(Z_\theta)| + |E(Z_\theta) - \hat{E}(Z_\theta)|$, which enables us to combine (3) and (4) into a single inequality for the divergence of Z_θ from its fitted value for given ξ. Solving this for ξ would give the single-use confidence region. More simply, thinking of a graph of Z against ξ for a limited region of ξ values, we may approximate the width of the confidence interval by 'height' divided by 'slope', essentially a local linear Taylor series expansion (see, for example, Carroll & Spiegelman 1986).

The approximate half-width is thus

$$\left\{ c_1 \sqrt{\left[\sum_1^{q'} \theta_j^2 \sigma_j^2 \right]} + c_2 \sqrt{\left[\sum_1^{q'} \theta_j^2 \sigma_j^2 s^2(\xi) \right]} \right\} \Bigg/ \left| \sum_1^{q'} \theta_j \hat{\beta}_j \right|. \tag{5}$$

This is the quantity we will minimize with respect to the q' channels with non-zero θ_j.

When σ_j^2 have to be estimated from the data they are replaced by the usual unbiased estimators $\hat{\sigma}_j^2$, but for this paper we have made no attempt to adjust the probability statements appropriately.

The non-zero θ_j which minimize (5) may be easily seen to be proportional to $\hat{\theta}_j$, where

$$\hat{\theta}_j = \left(\frac{\hat{\beta}_j}{\sigma_j^2}\right) \bigg/ \left(\sum_1^{q'} \frac{\hat{\beta}_j^2}{\sigma_j^2}\right), \tag{6}$$

$j = 1, \ldots, q'$, and for notational convenience we have assumed that the selected frequencies are the first q' out of q. Our estimator of ξ obtained from the linear compound of the $Z_j, j = 1, \ldots, q$, obtained from (6) is

$$\hat{\xi} = Z_{\hat{\theta}} - \hat{\alpha}_{\hat{\theta}}, \tag{7}$$

since $\hat{\beta}_{\hat{\theta}} = \Sigma \hat{\beta}_j \hat{\theta}_j, = 1$.

Coincidentally, (6) are the generalized least squares estimators conventionally used in multivariate controlled calibration, and represent a special unicomponent case of equation (2.16) of Brown (1982) with diagonal covariance structure. The minimized half-length of interval (5), substituted by (6), is

$$[c_1 + c_2\, s(\xi)] \bigg/ \sqrt{\left[\sum_1^{q'} \frac{\hat{\beta}_j^2}{\sigma_j^2}\right]}. \tag{8}$$

With c_1 and c_2 functions of q' and linked by $s(\xi)$ in the numerator of (8), our choice of q', and that subset of q' of the q channels, depends on the unknown ξ. However, if we choose equal probability levels $\gamma = \delta$, so that $c_1 = c_2 = c$, then minimization of (8) over q' becomes independent of ξ. Moreover, whether or not the levels are chosen equal, (8) is easily minimized by ordering the absolute values of the standardized slope coefficients $\hat{\beta}_j/\sigma_j$ and choosing the largest q' of these, and then taking that q' for which (8) is a minimum.

If we do not want the conditional interpretation for single-use curves characterized by two probability levels γ and δ, then inequalities (3) and (4) are replaced by

$$|Z_\theta - \Sigma\, \theta_j(\hat{\alpha}_j + \hat{\beta}_j \xi)| = |\Sigma\, \theta_j \sigma_j \epsilon_j^*| \leqslant \sqrt{[\Sigma\, \theta_j^2 \sigma_j^2 (1 + s^2(\xi)]} \sqrt{[\chi^2_{1-\eta}(q')]},$$

where $1-\eta$ is the single confidence level. Taking the same linearizing approximation to this gives a narrower interval but the same form of half-width and the same selection, provided $\eta = \gamma = \delta$.

Sufficient conditions for their to be little variation in the channels selected are as follows.

1. The variances of all the estimated regression slope coefficients $\{\hat{\beta}_j\}$ are small.
2. The slopes $\{\hat{\beta}_j\}$ belong to one of two non-empty sets. The first has absolute value far from zero, $|\beta_j| \gg 0$. The second group has slopes approximately equal to zero. For all slopes β_j and β_k in the first group it is assumed that the distances $\| \beta_j| - |\beta_k \|$ are large.

It is evident that the number of components selected is a monotonic decreasing function of the both γ and δ, so that if they are both chosen large enough only one component is selected. Typical values $\gamma = \delta = 0.1$ allow enough information to be retained in our detergent example.

In summary, our method orders the absolute values of $\hat{\beta}_j/\sigma_j$, with $\hat{\sigma}_j$ estimating σ_j when as usual the error standard deviation is unknown. It chooses the frequencies corresponding to the largest q' of these, where q' minimizes (8), and here c_1^2 and c_2^2 are tabulated χ^2 percentage points on q' degrees of freedom. Dependence on $s(\xi)$ disappears if the two confidence levels are equal.

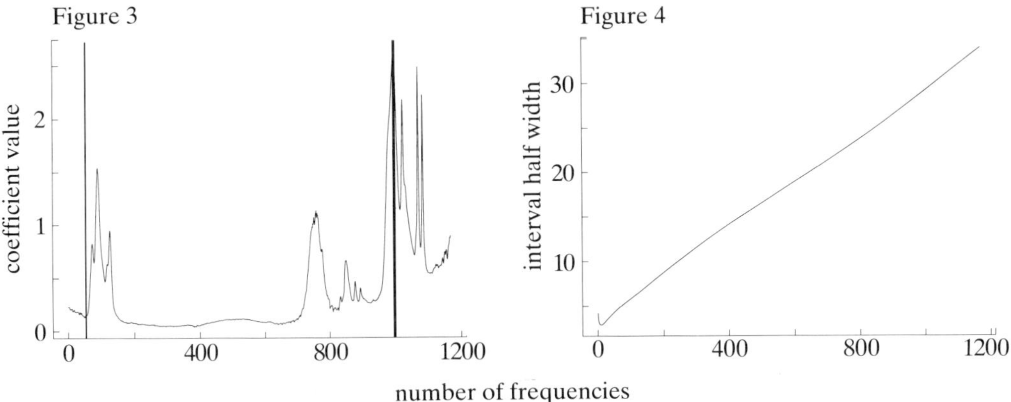

Figure 3. Component 1 multiple regression coefficients with selected frequency bands.

Figure 4. Component 1 confidence interval half-width by number of chosen frequencies. Minimum occurs at 11.

5. The detergent example

This data-set consists of absorptions at $q = 1168$ mid-IR, equally spaced frequencies (channels) in the range 3100 to 750 cm^{-1}. This detergent is a solution in water of four chemicals, with water making up the fifth component of the mixture. Only $n = 12$ carefully designed samples were available with the five percentages of ingredients recorded for each of these. The data are further described and analysed in Denham & Brown (1990), using the continuity of absorbance as a function of frequency. Since the calibration design was quite strictly controlled, we adopt a controlled calibration mode of analysis throughout, regressing absorptions on ingredients and then 'inverting' this relationship to predict a future composition from the absorptions for the sample. However, we have here concentrated on a single x-variable, whereas the data have five. We adopt the following approach to this. First, treating the data as five completely separate data-sets, we select sets of frequencies for each of these data-sets. We then are able either to continue to treat the five datasets as quite separate and predict each component from the inverted simple linear regressions, using (7); or to adopt a hybrid multiparameter approach involving (a) pooling the five sets of chosen frequency channels and (b) 'inverting' the $\Sigma q' = Q$ multiple regressions of absorption on components to predict each component in future from a set of absorptions. This inversion would itself be a weighted multiple regression of the Q-vector $Z_{(Q)}$ on vectors of regression coefficients with, as design matrix, the $Q \times 5$ matrix of regression coefficients and Q different variances from the earlier multiple regression of $Z_{(Q)}$ on components. The estimator is thus provided by (2.16) of Brown (1982), namely,

$$\hat{\xi} = (\hat{B}S^{-1}\hat{B}^{\mathrm{T}})^{-1}\hat{B}S^{-1}(Z_{(Q)} - \hat{\alpha}),$$

with the S the full error covariance matrix of that paper replaced by a diagonal $Q \times Q$ covariance matrix.

For component 1, figure 3, gives the bands of frequency channels selected for chosen probability levels $\gamma = \delta = 0.1$. The positions of the chosen frequencies are indicated by vertical lines, and are superimposed on the graph of slope coefficients for this component from multiple regression of absorbances on components. For this component there are 11 selected channels, in order {998, 999, 997, 996, 1000, 995,

Table 1. *Root mean squared prediction error*

method	component				
	1	2	3	4	5
simple LS	3.44	0.95	1.77	0.36	1.38
multiple LS	0.39	0.47	0.40	0.11	0.30
select simple LS	1.53	0.43	0.81	0.39	0.94
select multiple LS	0.14	0.21	0.15	0.11	0.23

54, 1001, 994, 53, 55}, making up two quite separate groups of frequencies, {53, 54, 55} and {994–1000}. Figure 4 is proportional to the half-width (8) as a function of q' for component one, depicting the minimum at $q' = 11$ and a sharp increase as q' increases, with the value at the minimum being around one tenth of that with all 1168 frequencies included. Figure 4 is typical of plots of (8) for the other components, and these have consequently been omitted, as have the other component plots paralleling figure 3. The number of frequencies selected for components two to five are 9, 3, 7, 17, and all the selected channels are quite distinct, so that the pooled set of frequencies for the hybrid multiparameter prediction involves $Q = 47$ selected channels.

Table 1 gives leave-one-out root mean squared prediction errors for the five components corresponding to four different methods. The tabulated results are thus cross-validated so that the selection of wavelengths is based on 11 observations and the 12th is predicted for every subset of size 11. Two of the methods incorporate all 1168 wavelengths and two involve selection of wavelengths as indicated above. The two methods are further dichotomized by whether they are uni- or multicomponent. As a consequence of the diagonal covariance structure, there is a substantial reduction of prediction error in using multicomponent methods, and within either uni- or multicomponent the selection procedures of this paper are beneficial for most components. The variance of the four detergent ingredient percentages were 16.4, 5.5, 4.9, 2.9, respectively, so that predictions are generally very accurate with very high percentages of variation explained. As a further reference point, partial least squares on all frequencies with four latent factors gave root mean square prediction errors of 0.20, 0.29, 0.15, 0.11, 0.18 for the four components and water. Although these are only marginally less good than our preferred last row of table 1, further work on the calibration of sugars in Brown (1991) shows partial least squares on all frequencies to lack robustness with respect to different prediction sets. After selection of frequencies by developments of the method of this paper, such non-robustness disappears.

Note that we have used mean squared error as a prediction criterion. For simple linear regression the mean squared error is infinite. However, the use of mean squared error is justified for $q \geqslant 3$, since it is finite if and only if the number q of frequencies is at least three (see Brown & Spiegelman 1991). This paper provides for the selection process and sharpens the results of Lieftinck-Koeijers (1988).

6. Commentary

We have provided a new method of channel frequency selection. The method is strictly applicable to unicomponent calibration, but we have demonstrated by example that a hybrid multicomponent method is also effective. This hybrid version coped with the multicomponents through $\hat{\sigma}_j$. If at a particular frequency other components also influenced the response, this would inflate the corresponding σ, reducing the standardized coefficient. It relies on discriminating between each component and the rest in turn. One can, however, envisage situations where the selection of frequencies should be based on a completely multicomponent method. This would be the case if for example particular frequencies were very good at discriminating between a pair of components on the one hand against the rest, but offered little discrimination between the pair, whose resolution could be achieved from other channels.

It is easy to see qualitatively how one might consider components simultaneously, and at the same time incorporate a correlation structure across frequencies. Some straightforward calculations following from Brown & Sundberg (1987) give the asymptotic, observed, second-derivatives, $p \times p$ matrix of the profile log-likelihood of ξ at the maximum likelihood estimate, that is, the information matrix for the vector of component ξs, as $\hat{B}\Gamma^{-1}\hat{B}^{\mathrm{T}}$. Here B is a $p \times q$ matrix of coefficients from the full version of model (1) formed by regressing the q absorptions on the $p = 4$ components. It is of interest to note that, when $p = 1$ and the $q \times q$ covariance matrix Γ is diagonal, then our criterion just amounts to accumulating the q' highest information components. Thus the ordering based on $\hat{\beta}_j^2/\sigma_j^2$ is quite natural. The use of information in itself does not give one a stopping rule which adequately takes account of the dimensionality of estimation.

It probably requires more experience with other data-sets to see whether or not the confidence levels γ and δ, can be prespecified and yet retain sufficient information. In the application of the methodology to the calibration of three sugars (Brown 1991), $\gamma = \delta = 0.001$ was chosen by cross-validation. That report also contains a simple modification to deal with an autoregressive error structure.

We are grateful for comments by Dr Rolf Sundberg on an earlier version, which led to an improved presentation. P.J.B. and M.C.D. are grateful to the SERC for providing a grant under the Complex Stochastic Systems Initiative. Shell UK also provided funding for work on selection in calibration with many variables. C.H.S. was funded by the US National Science Foundation and the Shell Development company.

References

Berry, D. A. 1988 Multiple comparisons, multiple tests and data dredging: a Bayesian perspective. In *Bayesian statistics* 3 (ed. J. M. Bernardo, M. H. DeGroot, D. V. Lindley & A. F. M. Smith), pp. 79–94. Oxford: Clarendon Press.

Brown, P. J. 1982 Multivariate calibration (with discussion). *Jl R. statist. Soc.* B, **44**, 287–321.

Brown, P. J. 1990 Partial least squares in perspective. *Analyt. Proc. R. Soc. Chem.* 303–306.

Brown, P. J. 1991 Variable selection with more variables than observations. Report to Shell Research, Sittingbourne.

Brown, P. J. & Mäkeläinen, T. 1992 Regression, sequenced measurements and coherent calibration. In *Bayesian statistics* 4 (ed. J. M. Bernardo, J. Berger, A. P. Dawid & A. F. M. Smith). Oxford University Press. (In the press.)

Brown, P. J. & Spiegelman, C. H. 1991 Mean squared error and selection in multivariate calibration. *Statist. Prob. Lett.* **12**, 157–159.

Brown, P. J. & Sundberg, R. 1987 Confidence and conflict in multivariate calibration. *Jl R. statist. Soc.* B **49**, 46–57.

Carroll, R. J. & Spiegelman, C. H. 1986 The effect of ignoring small measuring errors in precision instrument calibration. *J. Quality Control* **18**, 170–173.

Carroll, R. J., Spiegelman, C. H. & Sacks, J. 1988. A quick and easy multiple-use calibration-curve procedure. *Technometrics* **30**, 137–141.

Cross, A. D. & Jones, R. A. 1969 *An introduction to practical infra-red spectroscopy*, 3rd edn. London: Butterworth.

Dawid, A. P. 1988 The infinite regress and its conjugate analysis. In *Bayesian statistics* 3 (ed. J. M. Bernardo, M. H. DeGroot, D. V. Lindley & A. F. M. Smith), pp. 95–110. Oxford University Press.

Denham, M. C. & Brown, P. J. 1990 Calibration with many variables. Technical Report, Liverpool University.

Eisenhart, C. 1939 The interpretation of certain regression methods and their use in biological and industrial research. *Ann. math. Statist.* **10**, 162–186.

Fearn, T. 1983 A misuse of ridge regression in the calibration of a near infrared reflectance instrument. *Appl. Statist.* **32**, 73–79.

Lieftinck-Koeijers, C. A. J. 1988 Multivariate calibration: a generalisation of the classical estimator. *J. Multivariate Analysis* **25**, 31–44.

Mäkeläinen, T. & Brown, P. J. 1988 Coherent priors for ordered regressions. In *Bayesian statistics* 3 (ed. J. M. Bernardo, M. H. DeGroot, D. V. Lindley & A. F. M. Smith), pp. 677–696. Oxford University Press.

Marquardt, D. W. 1970 Generalised inverses, ridge regression, biased linear estimation and nonlinear estimation. *Technometrics* **12**, 591–612.

Martens, H. & Naes, T. 1989 *Multivariate calibration*. Chichester: Wiley.

Miller, A. 1990 *Subset selection in regression*. London: Chapman and Hall.

Osborne, B. G., Fearn, T., Miller, A. R. & Douglas, S. 1984 Application of near infrared reflectance spectroscopy to compositional analysis of biscuits and biscuit doughs. *J. Sci. Food Agric.* **35**, 99–105.

Scheffé, H. 1973 A statistical theory of calibration. *Ann. Statist.* **1**, 1–37.

Spezzaferri, F. 1985 A note on multivariate calibration experiments. *Biometrics* **41**, 267–272.

Stone, M. 1974 Cross-validatory choice and assessment of statistical predictions. *Jl R. statist. Soc.* B **36**, 111–147.

Stone, M. & Brooks, R. J. 1990 Continuum regression: cross-validated sequentially constructed prediction embracing ordinary least squares, partial least squares and principal components regression. *Jl R. statist. Soc.* B **52**, 237–269.

Sundberg, R. & Brown, P. J. 1989 Multivariate calibration with more variables than observations. *Technometrics* **31**, 365–371.

Williams, E. J. 1969 Regression methods in calibration problems. In *Bull. Int. Statist. Inst.*, vol. 43, pp. 17–28.

Global and local priors, and the location of lesions using gamma-camera imagery

By R. G. Aykroyd† and P. J. Green
Department of Mathematics, University of Bristol, Bristol BS8 1TW, U.K.

After a brief review of the paradigm of bayesian image restoration, we pose the question: If high-level prior information is available and usable, what is lost by modelling at the pixel level instead? Our discussion is based on a real application where this question is relevant: the use of gamma-camera imagery in the location of lesions. Procedures using 'global' prior information in the form of a structural model for the image are compared with those using more conventional 'local' priors, modelling only interactions among neighbouring pixels. We address in detail physical modelling and algorithms, and present results with both real and simulated data.

1. Introduction

Most modern work on statistical approaches to the extraction of information from digital images is based on the paradigm of bayesian image analysis pioneered by Besag (1983, 1986) and Geman & Geman (1984). This aims to embrace a wide variety of image analysis tasks, arising in applications right across the applied sciences from geography to medicine, within a framework based on probabilistic modelling and statistical inference. The essence of this framework is as follows.

1. The observed image or *record*, a finite array of numbers representing the pixellated signals or intensities, is regarded as a realization of a random vector.
2. The true 'state of nature', about which the record provides partial information, is a realization of a random array (or function) called the true image or *truth*.
3. Information about the truth available *prior* to observation is represented by a probability distribution.
4. The process generating the record, typically incorporating various forms of degradation such as blur, noise, geometrical distortion and discretization, is represented by the record's conditional distribution given the truth.
5. It is required to make inference about the truth, or some function thereof: this will be based on the conditional distribution of the truth given the record.

We will use the symbols x and y to denote the truth and the record respectively. All probability distributions will be expressed as densities with respect to appropriate measures, and will be denoted generically by p: thus $p(x)$, $p(y \mid x)$ and $p(x \mid y)$ represent the three distributions mentioned above.

Within this broad framework, there is considerable flexibility: we briefly elaborate on each of the items above. Even what constitutes the raw data y may be ambiguous when the image is recorded by modern instrumentation incorporating data

† Present address: Department of Mathematics, University of Bradford, Bradford BD7 1DP, U.K.

compression and pre-processing. The truth x may represent some physical reality such as the concentration of a chemical in human tissue, or the variety of a crop growing at a particular point in a remotely sensed scene, or it may simply represent an idealized version of the observed image y: the image that would be obtained by a perfect sensor.

The modelling of $p(x)$ may be based on a physical stochastic mechanism generating variation in x over replications, as might be appropriate in an industrial inspection application for example, or it may be a conventional representation of the form of variation that is expected. In some applications, the representation of information about x in probability terms is completely artificial, and the methodology used is really that of penalized likelihood. (In maximum penalized likelihood estimation, instead of maximizing the log-likelihood $\ln p(y \mid x)$ alone, one maximizes a penalized version, $\ln p(y \mid x) - J(x)$, say. This is formally equivalent to estimating by the posterior mode, when using a prior $p(x)$ proportional to $\exp(-J(x))$. For a review of penalized likelihood in regression problems (see Green 1987).) Whatever the basis for $p(x)$, it is important that this component be included. The dimension of x is usually comparable with or greater than that of y, so that some form of regularization is necessary in order to render inferences about x from y well-defined and correct. In this paper, we shall adopt the bayesian terminology without further comment.

The degradation model, $p(y \mid x)$, is often mechanistic in nature, aiming to capture the image production through an understanding of the physical processes involved, but it may be simply empirical.

Finally, although in this model-based scenario the posterior distribution is completely specified in principle, the large dimensionality of the arrays involved means that computational considerations may influence the specific inferential statements that may be made in practice. For example, one functional of the posterior distribution may be computed much more cheaply than another that would otherwise have been preferred, or approximations may be necessary.

The past six years or so have seen a number of successful applications of the framework described above, and also spin-offs of the same ideas into different areas, such as general multi-parameter bayesian inference (Gelfand & Smith 1990; Smith 1991), pedigrees in genetics (Sheehan & Thomas 1991), and geographical epidemiology (Besag *et al.* 1991). On the other hand, many traditional aspects of image processing technology have been barely touched by this statistical approach, and continue to make use of standard, mostly linear, techniques including spatial filtering, frequency domain methods, etc. Where traditional methods provide satisfactory results, the statistical approach may still have something to add, in that it offers the prospect of more complete inferential statements about the truth, beyond simply provision of a 'cleaned-up image': these include estimates of error, confidence bounds, tests of hypotheses about anomalies, and go on to higher level vision tasks such as probabilistic diagnosis and object recognition. Within the more mundane realm of image restoration and reconstruction, probabilistic methods seem to have had most success in applications where there is substantial stochastic noise, especially where this is not gaussian, where there is awkward geometrical degradation in addition to noise (as in the case of tomographic reconstruction, for instance), and where estimation of parameters in the models is also an issue.

One general field in which these ingredients are often present is that of medical imaging, particularly those techniques where image formation is a result of the counting of individual photons, such as emission tomography and other 'nuclear medicine' methods. This paper will focus on a particular problem arising in this area, that of locating lesions in organs from gamma camera imagery: this application will be introduced in the next section.

Considerable research effort in bayesian image analysis has been to do with the construction of prior distributions $p(x)$. From one perspective, this is just the general problem of modelling spatial stochastic processes in a tractable and flexible way, with parsimonious parametrization. Since we are interested in spatially discrete processes on regular lattices, the seminal work of Besag (1974) on lattice processes has been extremely important; here processes are specified as conditional autoregressions, that is, defined by the local conditional distribution of the value of x at each pixel i given the values at all other pixels, $p(x_i \mid x_{S\backslash i})$ say. If, for each i, this distribution is a function only of x_j for j in a set of pixels spatially close to i (that we denote by ∂i and term neighbours of i), then the resulting distribution is a Markov random field with respect to this neighbour structure. The local conditional probabilities must satisfy consistency conditions that are captured in the Hammersley–Clifford Theorem, which plays a central role by identifying positive Markov random fields precisely as Gibbs distributions.

Much of the modern statistical literature on image analysis has used such processes as prior distributions for the true image, not from principle, but for the thoroughly practical reason that most of the restoration and reconstruction algorithms based on posterior image distributions $p(x \mid y)$ can only be implemented efficiently when the prior is Markov with respect to a fairly sparse graph. For example, approaches using stochastic relaxation (e.g. the Gibbs sampler (Geman & Geman 1984)) and coordinate-wise maximization (e.g. iterated conditional modes (Besag 1986)) use the local conditional *posterior* distributions

$$p(x_i \mid x_{S\backslash i}, y),$$

while methods based on modifying the EM algorithm (see, for example, Green 1990) require the partial derivatives $\partial/\partial x_i \ln p(x)$. It is only with Markov fields $p(x)$ that these expressions can be computed very rapidly.

Although such prior distributions are important for practical reasons, it is clear that they can only to a very limited extent represent real prior information about the truth. Even with additional structure such as 'edge sites' between the pixels to help to organize boundaries in spatial classification problems (Geman & Geman 1984) or hierarchical structure such as that given by allowing the interaction parameter to vary stochastically in a spatially coherent way (see, for example, Clifford 1986), these remain *low-level* or local priors, capturing only relations involving small sets of pixels, such as continuity, gradients, boundary curvature, and so forth. In this sense they are a stochastic analogue of the penalty functionals that are used in non-parametric regression to fit regression relations assuming only smoothness.

Now, however, suppose there really is *high-level* or global prior information about the truth x available. Three pressing questions arise.

(*a*) Can this information be represented in probabilistic form?

(*b*) Can practical algorithms be constructed to extract information about the resulting posterior distribution $p(x \mid y)$?

(*c*) What is lost by not doing so, and using a low-level prior instead?

The answer to (*a*) is very context-specific, of course, and greatly influences the answer to (*b*). Much recent work in statistical image analysis seems to built on the implicit belief that the answer to (*c*) is 'not much'. This is expressed in various ways:

through arguments that properties of posterior distributions or restoration algorithms are sensitive only to *local* properties of the prior, through informal appeal to Savage's principle of precise measurement, and through special-case arguments such as that of Green (1986) in a situation where certain global structure is almost irrelevant to pixel-wise restoration algorithms.

Our principal aim in this paper is to examine a situation where the answers to both (*a*) and (*b*) are 'yes', and where, through some extensive experimentation, we can provide an answer to (*c*). The context is the use of gamma-camera imagery for the detection and location of tumours. The models we use are slightly idealized, but we believe that the comparisons we provide between use of local and global priors in bayesian image analysis have more general relevance. In the next section, we describe the process of image capture using the gamma camera. Section 3 includes a discussion of prior modelling of tumours in this context, and both local and global priors are defined. In §4 we describe our methodology, and in §5 we report on some of our experiments in the use of the resulting techniques, initially on artificial examples, and then on some real gamma-camera data.

For completeness, we close this introduction by mentioning some other alternatives to pixel-based priors that have been discussed. Recent work that comes quite close in spirit to the present paper is that of Dupuis *et al.* (1991) concerning optical astronomy data obtained from the defective Hubble space telescope. Here the authors are concerned with the classification of astronomical objects, such as binary stars, of simple basic form, so that realistic parametric models can be constructed. Avoiding pixellation altogether in modelling the truth can be accomplished by use of appropriate sets of basis functions. For example, if trigonometrical polynomials were used, this would amount to discretization in the frequency domain instead of the original space. Jones & Silverman (1989) use Zernike and Chebyshev polynomials in an orthogonal series approach to the PET reconstruction problem. When the true image is of known, highly structured form, but uncertain in terms of fine detail, the approach of Chow *et al.* (1988) and Amit *et al.* (1991) is attractive: here the stochastic aspect of the prior is provided by representing the truth as the result of a random deformation of a fixed ideal template. Ripley & Sutherland (1990) use related ideas in modelling spiral structure in galaxies.

The underlying effect of all of the approaches just discussed is a drastic reduction in the number of degrees of freedom of the random elements in the true image model from the very large number inevitable if one works at the pixel level. This sounds attractive, but what is the cost in lost generality and robustness? Which approach, global or local, gives better performance when computational effort is also taken into account?

2. Gamma-camera imagery

Gamma-camera imaging is a modern medical diagnostic technique aimed at studying function rather than form. The patient is injected with, or inhales, an appropriate drug or some other substance that is known to become concentrated in the organ of interest, in a manner related to the phenomenon under study, for example, metabolic activity. Before injection, the substance is labelled with a suitable radio-isotope: photon emission therefore occurs in the organ, at a rate varying spatially according to the local concentration of the substance. Indirect measurements of this concentration can thus be made from counts of emitted particles collected in appropriate detectors.

The physical and operational details of the gamma camera are described in the monograph of Larsson (1980). In essence it consists of a parallel-bore collimator fronting a fluorescing crystal. Incoming photons successfully traversing the narrow tubes of the lead collimator strike the crystal, causing a fluorescence, whose location is measured by an array of photo-multiplier tubes behind the crystal. The circuitry and software of the system record these locations, applying spatial discretization on to a square grid. The whole device is mounted on a ring in a vertical plane so that it can be swung round the patient, who lies on a horizontal couch, and thus used to collect tomographic information: this is the methodology of single photon emission computed tomography (SPECT). A bayesian approach to reconstruction from SPECT data has been investigated by Geman & McClure (1987) and Green (1990). The gamma camera is also routinely used to create sequences of images, to allow the study of dynamic processes such as the flow of material through the kidneys, or the function of the beating heart.

In this paper, however, we are concerned with the most fundamental use of the device, that of making a single, direct, two-dimensional image representing a projection onto the plane of the camera of the isotope concentration within the patient. The recorded image y thus consists of a finite array, usually square, of counts of detected photons, which we will index as $\{y_t, t \in T\}$. The truth x in this case is the spatial distribution of the isotope: let $\{x(\boldsymbol{u}), \boldsymbol{u} \in \mathbb{R}^3\}$ denote the isotope concentration at spatial coordinates $\boldsymbol{u}$ with respect to appropriately chosen axes. We need to model $p(y \mid x)$, the joint probability function of $\{y_t, t \in T\}$ given x. In one sense, this is entirely straightforward: to a close approximation, photons are emitted in an inhomogeneous space-time Poisson process, with rate proportional to $x(\boldsymbol{u})$, and each photon taking an independent, completely random direction of flight. Through nuclear interactions with other particles, both within the body and elsewhere, that lead to scattering and absorption, most of the photons are lost to the system. Some small fraction arrive at the collimator with paths that are sufficiently close to perpendicular to its face that they pass through and are detected. We neglect 'dead-time' effects in detection, so that we assume all photons striking the crystal are recorded. Each photon is recorded at most once, so that we are observing independent, superimposed, thinned Poisson processes of arrival at each detector, and thus the model

$$y_t \sim \text{Poisson}\left(\int a_t(\boldsymbol{u})\, x(\boldsymbol{u})\, \mathrm{d}\boldsymbol{u}\right),$$

independently for all $t \in T$, is derived. Here $a_t(\boldsymbol{u})$ is the mean rate of arrival at detector t from emission at unit rate from a point source of unit concentration at $\boldsymbol{u}$. These functions express the physical factors of absorption and scattering, and the geometrical relation between $\boldsymbol{u}$ and t, including the traversal of the collimator bores. This statistical model for nuclear medicine data is similar to that derived by Shepp & Vardi (1982).

In any reconstruction method based on this model, values for these functions $\{a_t(\boldsymbol{u})\}$ will be needed, and these may be obtained either by associated experiments with known sources, or by appropriate physical modelling. An example of the latter approach is given in some detail in Green (1990) for the case of SPECT.

We are concerned here, however, with the straightforward use of the gamma camera where a single direct view is obtained, so that the integral $\int a_t(\boldsymbol{u})\, x(\boldsymbol{u})\, \mathrm{d}\boldsymbol{u}$ represents a pixellated, blurred, attenuated projection. As a result of the narrow angle of view due to the collimator, there is very little information in these projections about the third dimension of x, orthogonal to this projection. To be more precise, suppose that the u_3 axis represents that orthogonal direction, oriented into the patient's body, that we can approximate the skin surface nearest the camera by the (u_1, u_2) plane, and that the coefficient of linear attenuation within the body is a constant μ. Then, if detector t has coordinates (t_1, t_2, t_3) with respect to the same axes, we will have approximately

$$a_t(\boldsymbol{u}) = \mathrm{e}^{-\mu u_3} h(u_1 - t_1, u_2 - t_2),$$

for an appropriate point-spread function, h, that will be assumed known. Thus

$$\int a_t(\boldsymbol{u})\, x(\boldsymbol{u})\, \mathrm{d}\boldsymbol{u} = \int h(u_1 - t_1, u_2 - t_2)\, x^*(u_1, u_2)\, \mathrm{d}u_1\, \mathrm{d}u_2,$$

where

$$x^*(u_1, u_2) = \int_0^\infty x(u_1, u_2, u_3)\, \mathrm{e}^{-\mu u_3}\, \mathrm{d}u_3.$$

This expresses the blur, attenuation and projection. Finally, we will discretize x spatially onto the same grid as that on which the data are recorded, and obtain a model in the form

$$y_t \sim \mathrm{Poisson}\left(\sum_s h_{ts} x_s\right), \tag{1}$$

where x_s is the discretized version of the attenuated projection $x^*(u_1, u_2)$, and h_{ts} is the discretized point spread function. Other approaches to the discretization would be possible, of course: Jennison & Jubb (1991) discuss restoration on multiple pixel scales, including some much finer than that corresponding to the data, or indeed we could work in continuous space by using a set of smooth basis functions.

3. Priors for ellipsoidal lesions

In this section we discuss both global and local prior distributions that we will use to model a lesion in the organ being imaged. We suppose that it is known that exactly one lesion is present, and the problem is to locate it. There are fairly obvious extensions, which we have not pursued, to problems of detection, and problems involving several lesions: one can simply place a prior distribution over the number of lesions, including zero.

The simplest credible model for a lesion is to regard it as ellipsoidal in shape with a constant isotope density, surrounded by healthy tissue, again with a constant but higher density. In ideal imagery the lesion would thus be seen as an elliptical 'cold spot'. If the organ is reasonably regular in shape, then, in the absence of a lesion, the attenuated projected density x_s would be a smooth function, which we will regard as locally planar. The ellipsoidal lesion will introduce a depression into the surface, which, if its diameter is small compared with the attenuation length constant μ^{-1}, will be hemi-ellipsoidal in profile. Thus, before blur, pixellation and noise, the ideal image surface will be, locally at least, a ramp with a 'hole' of known shape.

Such an intensity surface can be characterized very economically, with three parameters to specify the ramp, two for the hole position, one for its depth and three for its size and shape. Our global prior will be the model that takes the truth to have this ideal form, with a uniform prior distribution over each of the ramp and hole location parameters. In this study, we assume the hole depth, size and shape to be known. The approach should extend readily to deal with the general case, where informative prior distributions for the additional parameters would usually be appropriate, but that is beyond the scope of this paper.

Two pixel-based priors will also be considered, for comparison. One is an 'implicit discontinuity' model similar to that introduced by Geman & McClure (1987) for use in SPECT:

$$p(x) \propto \exp\left(-\beta \sum_{s \sim r} w_{sr}\, \phi\left(\frac{x_s - x_r}{\delta}\right)\right), \tag{2}$$

where β and δ are parameters, the sum is over pairs of neighbours in an 'eight nearest neighbour' system on the square lattice, and the weights w_{sr} are 1 for orthogonal and $1/\sqrt{2}$ for diagonal neighbours.

If ϕ were a positive definite quadratic function, this would represent a gaussian random field: our experience, agreeing with that of others, is that discontinuities in the image surface are excessively smoothed out in restorations based on a gaussian prior. What is needed is a function that increases less rapidly than $\phi(u) = u^2$ for large (absolute) values of its argument: we use $\phi(u) = \operatorname{logcosh}(u)$, as in Green (1990), which has this property, yet for which $p(x)$ remains log-concave.

The second pixel-based prior we consider is a simple modification to this, that gives equal maximum probability to all planar image surfaces, not just the constant ones. As suggested, for example, by Geman (1991), we replace the difference in (2) by a laplacian, so that the prior has the form

$$p(x) \propto \exp\left(-\beta \sum_{s} \phi\left(\frac{x_s - \bar{x}_s)}{\delta}\right)\right), \tag{3}$$

where $\bar{x}_s$ is the mean of the four nearest neighbours. For pixels at the edge of the image, with fewer than four orthogonal neighbours within the image, we replace $\bar{x}_s$ in expression (3) by the value at s of the least squares plane fitted to those of the eight nearest neighbours to s on the lattice that are within the image, and scale δ so as to standardize the variance of all resulting arguments of ϕ. Effectively, (3) penalizes discontinuities in gradient whereas (2) penalizes discontinuities in level.

Both (2) and (3) are chosen with the ideal ramp-plus-ellipsoidal-hole model in mind, and might be expected not to smooth out the abrupt changes in gradient at the perimeter of the hole; however, they only reflect 'local knowledge' and in no sense represent the complete profile of the image surface. On the other hand, their very lack of specificity to the lesion model suggests robustness to departures from this ideal form, which would clearly be a desirable property in handling real images. Further, there is more accumulated expertise in using such low-level models, and the prospects of general purpose software being developed to deal with them.

4. Methodology and algorithms

We are here aiming at Bayes inference using a fully specified model, but may have to accept compromises, forced by computational considerations, over what aspects of the posterior distribution are extracted.

The methods for image restoration and lesion location described here are based on approximate posterior distributions calculated using various forms of the Metropolis algorithm. This is a dynamic Monte Carlo technique, in which an ergodic image-valued Markov chain is constructed that has the posterior distribution of interest as its limiting distribution. For discussion see Hammersley & Handscomb (1964), Hastings (1970), Geman (1991) or Green & Han (1991).

We first consider this in detail for the case of the global prior. The methodology has been implemented at present only for the case where the hole in the true image surface x caused by the lesion has a known profile (depth, size and shape). There remain only five degrees of freedom to describe the surface (intercept, two gradients and two hole location coordinates). These reduce to three in the case of a one-dimensional image, a situation studied in greater detail in our simulation experiments.

The Metropolis algorithm is used in the following manner for this problem (Hastings (1970) is a convenient general reference). Each of the five (or three) parameters is considered in turn. A proposed new value for the parameter is drawn from a Normal distribution centred at the current value. It is valid to use any positive value τ^2 for the variance of this proposal distribution; however, the rate of convergence to equilibrium is greatly influenced by τ^2. This issue is discussed in detail by Green & Han (1991). In the present context, separate experiments suggested that the best performance was obtained with τ^2 somewhat larger than the variance of a gaussian approximation to the Gibbs sampler for this problem. Let the true discretized image surface corresponding to the current parameter values be x, and that obtained with the proposed change be x'. The proposal is accepted, and the parameter value updated accordingly with probability $\min\{1, p(x' \mid y)/p(x \mid y)\}$. Otherwise, it is rejected, and no change is made. The expression for the acceptance probability reduces to $\min\{1, p(y \mid x')/p(y \mid x)\}$ since the prior is uniform, and further simplifies considerably for numerical calculation when the point-spread function h has limited width. Each of the parameters is considered, in the same way, and the whole cycle repeated until stability is apparent.

Convergence is assessed by monitoring the values of several one-dimensional functionals of the evolving process, including the total posterior energy, the relative frequency of acceptance, and parameters of a least-squares plane fitted to the image surface. The stochastic processes thus defined are plotted against number of sweeps, and a judgement made as to how many iterations must be discarded to be confident that the initial transient is effectively finished. A full discussion of this difficult issue is not appropriate here: the reader is referred to the excellent lecture notes on the subject by Sokal (1989). It is clear that there can be no absolute guarantee that convergence has been achieved, given a finite realization. The process is continued for several hundred more sweeps, accumulating the sample means of the evolving parameter values. Our estimate of x is the surface corresponding to these mean values.

Turning now to the pixel-based priors, a similar procedure is followed, except that now it is individual pixel values that are considered in turn. Again Normal distributions centred at current values, with variances chosen in a similar way, are used to generate proposals. The acceptance probability has a more complicated form in the presence of a non-trivial prior: it reduces to

$$\min\left\{1, \frac{p(y \mid x')}{p(y \mid x)} \frac{p(x_i' \mid x_{\partial i})}{p(x_i \mid x_{\partial i})}\right\} \tag{4}$$

when pixel i is being considered. The last factor involves computing only a few values of the function ϕ, whether the pixel difference prior (2) or the Laplacian based prior

(3) is used. After discarding the initial transient, we estimate the posterior expectations of pixel values x_s by accumulating sample means.

Noise and blur are reduced or removed by this process, but of course the result cannot be expected closely to approximate the idealized ramp-plus-ellipsoidal-hole model. This will always be true when using a pixel-based prior, even in the case of artificial data generated from the model, in the presence of noise of the magnitude we are dealing with (the Poisson means are in the range (20, 100)). In our experience, the position of the hole is always well-defined visually in a display of the estimated posterior mean true image, which provides a clean and acceptable function estimate.

For some purposes in practice, however, a parametric point estimate of lesion location would be needed. This will also allow us here to make a fair comparison in numerical terms with the approach using the global prior. Thus some post-processing of the posterior mean is required to extract numerical estimates of the parameters of the surface. The algorithm we use is simplistic, and it is certainly quite inadequate as an estimator of lesion location if applied directly to the raw data. We subtract our estimated posterior mean truth from an over-smoothed version, that reflects the general trend of the surface, but smooths out any holes. Such a surface can be obtained by any convenient method: for example, we used a cubic spline smooth of the posterior mean in our one-dimensional experiments, and another bayesian estimate, with much greater β, in two dimensions. The difference image consists of a pattern of positive and negative lobes: the centroid of the highest positive lobe provides a cheaply computed estimate of hole location, that proves experimentally to be fairly robust to smooth departures from the ramp-plus-hole model. With the lesion located, the remainder of the surface is estimated by a least squares plane fitted to the posterior mean with the hole profile subtracted. This proposed procedure is arbitrary, but the estimates it produces are stable and robust to perturbations in the method. It is extremely cheap computationally compared with any formal method of fitting the ramp-plus-hole model, which would require searching or iterating as the model is severely nonlinear.

As in any application of a stochastic algorithm to solve a deterministic problem, it is appropriate to attempt to quantify the errors introduced into the values of our estimates by the algorithm. We should generally be satisfied that sufficient precision was being achieved if such 'Monte Carlo variances' did not exceed, say, 1% of the sampling variances of the estimators. Estimation of Monte Carlo variance is discussed by Sokal (1989) and Green & Han (1991). The truncated periodogram estimator advocated there is not appropriate in the present context because the functionals of the process of interest, the parameter estimates extracted by post-processing, are nonlinear. An estimator obtained from 'blocking' the Monte Carlo realization, related to a proposal by Hastings (1970), is more easily used. Suppose that the Metropolis sampler is run for $N = bk$ sweeps after the initial transient has been discarded, so that it may be assumed the process is an equilibrium. Let $\bar{x}_{k,i}$ be the pixel-wise mean of the realized value of x, over sweeps $(i-1)\,k+1$ to ik inclusive, for $i = 1, 2, 3, \ldots, b$. This is the estimated posterior mean from the ith of b blocks of k consecutive sweeps: $\bar{x}_{N,1}$ is the mean over the whole run. Let the post-processing used to extract an estimate of a particular parameter be represented by the function f. Then if $s^2_{b,k}$ is the sample variance of the post-processed block means:

$$s^2_{b,k} = \frac{1}{b-1} \sum_{i=1}^{b} \left[f(\bar{x}_{k,i}) - \frac{1}{b} \sum_{j=1}^{b} f(\bar{x}_{k,j}) \right]^2,$$

we will have, under weak conditions on the Markov chain being simulated (see Green & Han 1991),

$$E(s^2_{b,k}) \sim \operatorname{var}\{f(\bar{x}_{k,i})\}, \quad \operatorname{var}\{f(\bar{x}_{k,i})\} \sim c/k \quad \text{as } k \to \infty,$$

so it follows that $b^{-1} s^2_{b,k}$ is an approximately unbiased estimator of $\operatorname{var}\{f(\bar{x}_{N,1})\}$ as $k \to \infty$. For stability of this estimator, we will also require b to be large, so for fixed $N = bk$ there will be a trade-off between number of blocks and block length. A practical suggestion would be to compute the estimate of Monte Carlo variance for several block lengths, and to use the smallest blocks for which any trend has levelled out.

One important aspect of this methodology has not been addressed at all in this study. That is the choice of values for the hyperparameters β and δ appearing in (2) and (3). From a formal Bayes perspective, these parameters will either be known, or themselves subject to prior modelling. An empirical Bayes viewpoint would allow them to be estimated from the data, perhaps using Besag's pseudo-likelihood method as implemented for similar problems, not involving blur, by Qian & Titterington (1991*a*), or the Gibbs/EM approach used in Geman & McClure (1987). It should be recognized that this estimation problem is not straightforward, when taking account of the Poisson variation, the substantial blur, the fact that not only canonical parameters have to be estimated and the fact that the prior distribution will only be an approximate representation of reality. These difficulties are additional to those identified in the very useful discussion by Qian & Titterington (1991*b*). Further, good estimation is not the same as successful selection for the ultimate goal of lesion location. In this paper, values of β and δ are chosen, quite informally, to give good performance in practice. Clearly, further work would be needed on this issue before routine application of this methodology.

5. Experiments

We have conducted a rather extensive series of experiments using the models and methods described above. In this section, we report on some of these in two subsections, one concerning simulation experiments with one-dimensional images, and the other using real gamma-camera data.

(*a*) *One-dimensional simulations*

Artificial data from the model (1) were generated in order to examine the performance of our methods under repeated sampling. The truth was constructed using the ramp-plus-hole model in one dimension, in which the ramp had the form

$$x_s = \alpha_0 + \alpha_1 s$$

using values $\alpha_0 = 20$ and $\alpha_1 = 0.6$. The hole had a fixed profile, that of a semi-ellipse, with vertical semi-axis 20 and horizontal semi-axis 9.5, centred at location $\gamma = 60$. This surface was discretized into 128 pixels, not by sampling, but by integrating over 128 contiguous intervals $[0.5, 1.5]$, $[1.5, 2.5], \ldots, [127.5, 128.5]$. This seemed to be a better approximation of the effect of discretization in the real gamma camera. The point-spread function h_{ts} appearing in (1) was taken to be gaussian in shape, integrating to 1, with spread specified by its standard deviation σ; for numerical efficiency, the gaussian curve was truncated at $\pm 3\sigma$. (Point-spread functions are commonly parameterized in image-processing by the full width at half maximum

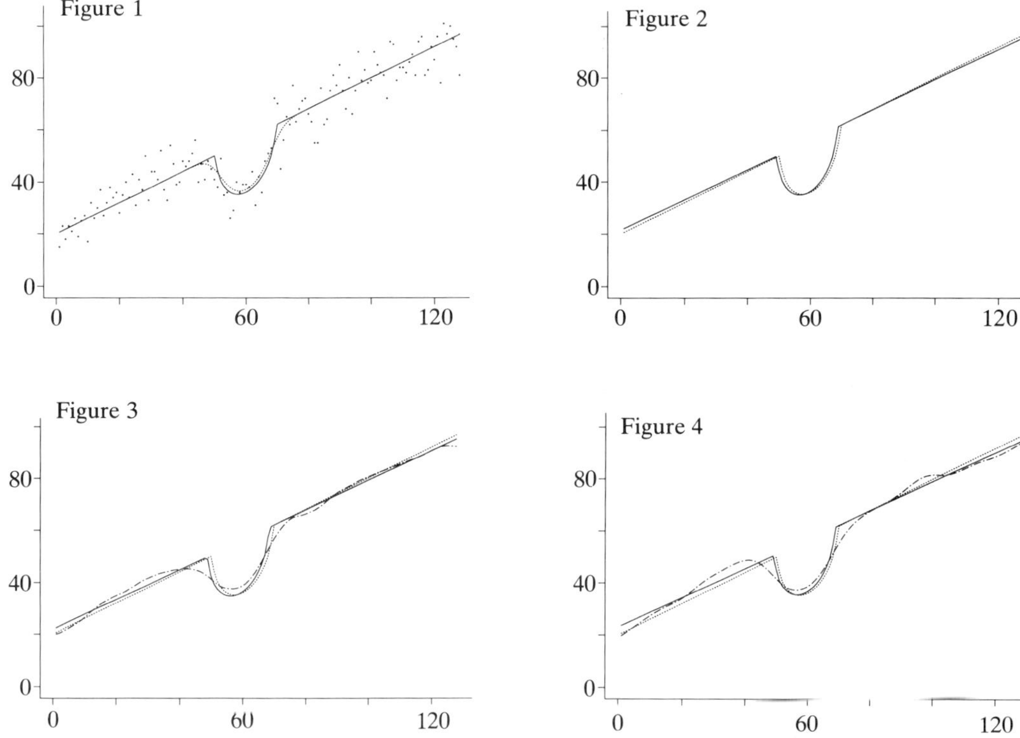

Figure 1. Truth, blurred truth and example data-set ($\sigma = 3$). ——, truth; ---, blurred truth; •, example data-set.

Figure 2. Reconstruction using global prior ($\sigma = 3$). ——, posterior mean; ---, truth.

Figure 3. Reconstruction using local difference prior ($\sigma = 3$). ——, post processed; ---, truth; —·—, posterior mean.

Figure 4. Reconstructions using local laplacian prior ($\sigma = 3$). ——, post processed; ---, truth; —·—, posterior mean.

(FWHM). For a gaussian function, FWHM is 2.35σ.) In our experiments σ was varied, taking values 0, 1, 3 and 5, and its value assumed known in the analysis. A modest complication is introduced by edge effects, which are quite pronounced when the blur is severe. Values of x_s needed from pixels s beyond the boundary of the observed image were replaced by values extrapolated by linear regression through values close to but within the boundary. (For efficiency, this was actually implemented by appropriately modifying h_{ts}.) The effect is that the blur leaves a linear ramp unaltered.

For each level of blur, 100 independent replicate data-sets were generated from the model (1), using an exact Poisson random number generator. The range of variation of the artificial data in this study was chosen to correspond broadly to that of the real gamma-camera imagery we have encountered.

In figure 1, we display the true ramp-plus-hole curve, the result after application of the gaussian point spread function with $\sigma = 3$, and one of the 100 corresponding data-sets. It is evident that both blur and noise make the problem of locating the hole considerably more difficult.

Each of the three methods proposed in the paper was applied to each data-set. In the case of the 'local' prior methods, values for β and δ were chosen by trial and error,

Table 1. *Mean parameter estimates from one-dimensional simulation experiments*

global prior (number of sweeps = 2000 (after discarding 500), $\tau^2 = \{4.0, 9.0\times10^{-4}, 4.0\}$)

σ	$\bar{\alpha}_0$	$\bar{\alpha}_1$	$\bar{\gamma}$	mse (l)	mse (α_0)	mse (α_1)	mse (γ)
0	19.85	0.6008	60.05	1.472	1.466	3.01×10^{-4}	0.422
1	19.84	0.6044	60.11	1.728	1.675	3.68×10^{-4}	0.500
3	19.90	0.6016	60.01	1.843	1.140	3.06×10^{-4}	0.775
5	20.07	0.6015	59.92	2.362	1.117	2.88×10^{-4}	1.395

local difference prior (number of sweeps = 5000 (after discarding 500), $\tau^2 = 25$)

σ	β	$\bar{\alpha}_0$	$\bar{\alpha}_1$	$\bar{\gamma}$	mse (l)	mse (α_0)	mse (α_1)	mse (γ)
0	3.0	20.40	0.5890	60.18	1.878	1.891	4.828×10^{-4}	0.560
1	2.5	20.27	0.5946	60.12	1.849	1.532	4.401×10^{-4}	0.716
3	2.0	20.17	0.5959	60.11	1.929	1.293	3.306×10^{-4}	0.941
5	1.5	20.29	0.5968	60.10	2.624	1.471	3.681×10^{-4}	1.631

local laplacian prior (number of sweeps = 5000 (after discarding 2000), $\tau^2 = 0.25$)

σ	β	$\bar{\alpha}_0$	$\bar{\alpha}_1$	$\bar{\gamma}$	mse (l)	mse (α_0)	mse (α_1)	mse (γ)
0	40	19.74	0.6031	60.01	1.844	1.771	4.104×10^{-4}	0.664
1	35	19.93	0.6030	60.01	2.172	1.784	5.190×10^{-4}	0.799
3	25	20.02	0.6011	60.00	2.620	1.602	4.655×10^{-4}	1.396
5	25	20.04	0.6009	59.92	2.886	1.521	3.927×10^{-4}	1.952

to give good average performance across all data-sets at a given level of blur. The values used for β are indicated in table 1: δ was fixed throughout at 2.0 for the pixel difference prior, and 0.5 for the laplacian prior. These values were chosen as giving best performance in terms of mean squared error. The table also indicates the values of τ^2 and the number of sweeps that were used. Figures 2, 3 and 4 display the results of our methods in graphical form, for the single data-set illustrated in figure 1. These figures show the estimated true images, obtained from the estimated posterior means as described in §4, and also the post-processed versions in the cases of the two 'local' methods.

In table 1, the performance across all levels of blur and all data-sets is summarized by a tabulation of means and mean-squared errors across replications. For a particular level of blur, and a particular method, let $\hat{\alpha}_0^{(r)}$, $\hat{\alpha}_1^{(r)}$ and $\hat{\gamma}^{(r)}$ be the parameter estimates for the rth data-set obtained from the posterior means, after post-processing if necessary. The columns headed $\bar{\alpha}_0$, $\bar{\alpha}_1$ and $\bar{\gamma}$ give the values of $\Sigma\,\hat{\alpha}_0^{(r)}/100$, etc. The columns headed mse (α_0), etc., give the mean-squared errors $\Sigma\,(\hat{\alpha}_0^{(r)}-\alpha_0)^2/100$, etc., and that headed mse (l) the average per-pixel mean-squared error

$$\sum_{r=1}^{100}\sum_{s=1}^{128}(\hat{x}_s^{(r)}-x_s)^2/12800,$$

where $\hat{x}_s^{(r)}$ is the estimated ramp-plus-hole surface based on $\hat{\alpha}_0^{(r)}$, $\hat{\alpha}_1^{(r)}$ and $\hat{\gamma}^{(r)}$. For the two local priors, the mean-squared errors for the posterior means before post-processing were almost exactly four times the mse (l) figures. From these means and mean-squared errors, bias and variance can readily be estimated. Since the replicates are independent, approximate t-statistics can be calculated to assess the significance of any apparent biases. While there is no significant evidence that the estimates

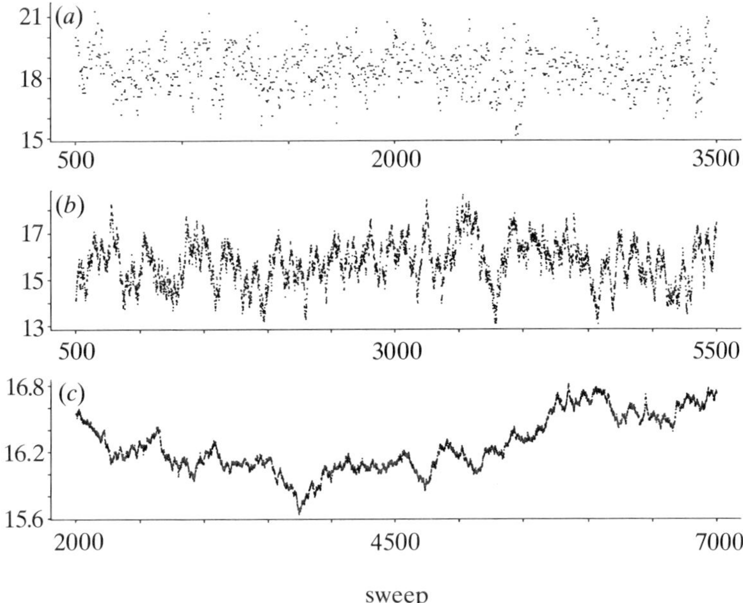

Figure 5. Monitoring statistics. (*a*) Global prior: intercept parameter. (*b*) Local difference prior: intercept of LS line. (*c*) Local laplacian prior: intercept of LS line.

based on the global and laplacian priors are biased, the method based on the pixel difference prior on average estimates the ramp as too shallow, with the hole position too far to the right.

In terms of overall mean-squared errors, however, all three methods are performing comparably well. Note the role of the blur parameter σ. As this increases, the errors in estimating γ increase while those of α_0 and α_1 reduce slightly.

A benchmark for the estimated sampling variances in this experiment is provided by those obtained by maximum likelihood estimation of (α_0, α_1) in fitting the ramp alone to the first of the stimulated data-sets, for example by Fisher scoring. The resulting variances were (0.968, 0.00272), slightly less than those obtained here.

In making firm comparisons between the methods, some caution is necessary. First, we need to take account of the computing times involved. For the numbers of sweeps indicated in table 1, the times for each of the three methods were between 3 min and 4 min for a single data-set in the case $\sigma = 0$ (all times quoted were obtained on a Sun Sparcstation 1). Thus the effort saved in only resampling three values each sweep instead of 128 is offset by the additional effort in repeatedly recomputing the ramp-plus-hole surface and all likelihood terms. However, as the blur increases to the rather excessive maximum value considered, $\sigma = 5$, while the computing times for the global method increase by a factor of about 2.5, those for the local methods multiply 14 times. Thus, from this perspective, the local methods become increasingly uncompetitive as blur increases.

More seriously, there is the question of whether convergence is really achieved. In this one-dimensional case, convergence of the pixel-based methods, especially the laplacian one, is very slow indeed. The parameter settings used here were selected after exhaustive experimentation as giving the fastest convergence available, yet plots of our monitoring statistics, such as those in figure 5*c*, suggest that autocorrelation times remain long. This difficulty does not apply to the global

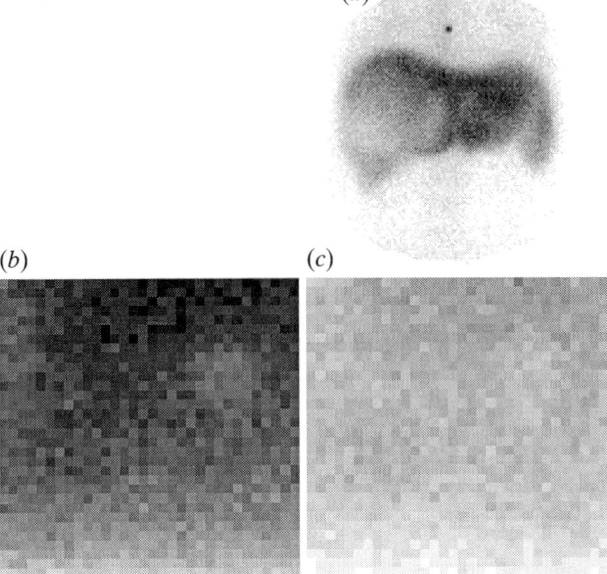

Figure 6. Gamma-camera data. (*a*) Image of liver. (*b*) Extract of phantom data. (*c*) Thinned phantom data.

Table 2. *Variances for global prior*

	estimate variances			Monte Carlo variances		
σ	$\bar{\alpha}_0$	$\bar{\alpha}_1$	$\bar{\gamma}$	$\bar{\alpha}_0$	$\bar{\alpha}_1$	$\bar{\gamma}$
0	1.442	3.004×10^{-3}	0.4222	0.010976	3.027×10^{-6}	0.56251×10^{-3}
1	1.186	3.081×10^{-3}	0.5021	0.005000	1.470×10^{-6}	0.88771×10^{-3}
3	1.128	3.035×10^{-3}	0.7750	0.006648	1.765×10^{-6}	1.47200×10^{-3}
5	1.113	2.859×10^{-3}	1.3955	0.011473	3.417×10^{-6}	2.04096×10^{-3}

method (see figure 5*a*), for which in table 2 we give Monte Carlo variances calculated by the blocking method. Comparing these with the sampling variances in the same table confirms that the Monte Carlo runs we used were more than adequate in length.

Our conjecture is that the slow convergence in the local methods is due to the one-dimensional nature of this example, a view which can be supported by heuristic arguments about connectivity. Notwithstanding this difficulty, we have seen that the methods do perform well; but we cannot confidently claim we are in fact evaluating the posterior means.

(*b*) *Gamma-camera phantom data*

To investigate the performance of gamma-camera imaging, and to calibrate equipment, medical physicists occasionally conduct experiments in which the patient is replaced by a physical 'phantom', usually a perspex box containing radioactive liquid, emitting photons. The data we use here were kindly provided by Dr C. J. Gibson, of the Northern Regional Medical Physics Department at the Dryburn Hospital, Durham.

Figure 6 displays three gamma-camera data-sets as grey-scale images. Panel (*a*) is an image of a patient's liver: the existence of a very large lesion is evident from visual

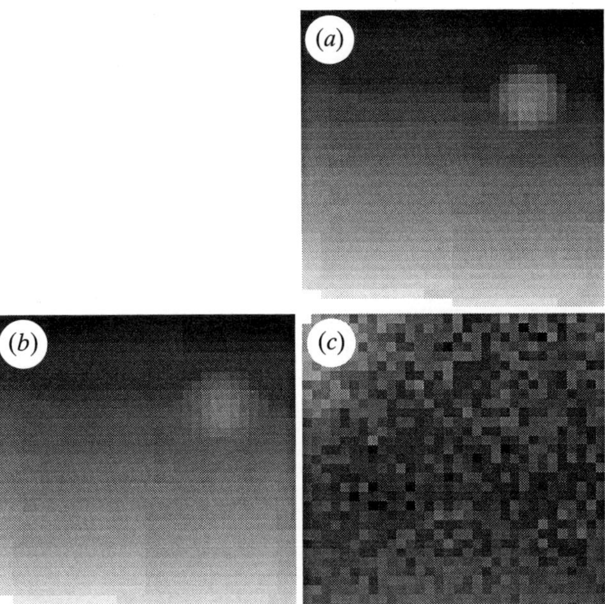

Figure 7. Reconstructions using global prior. (*a*) Posterior mean. (*b*) Fitted values. (*c*) Raw residuals.

Table 3. *Parameter estimates from two-dimensional examples*

(Phantom data. ($\delta = 30$. Number of sweeps = 2000 (after discarding 500). Global model, $\tau^2 = \{0.025, 0.00025, 0.01, 0.025, 0.025\}$; local difference, $\tau^2 = 500$; local laplacian, $\tau^2 = 180$.))

prior	β	α_0	α_1	α_2	γ_1	γ_2
global		48.88	0.4366	5.9674	24.67	22.93
local difference	1	48.51	0.6409	5.4295	24.18	22.85
local laplacian	10	48.08	0.6680	5.4402	24.33	22.78

(Thinned phantom data. ($\delta = 10$. Number of sweeps = 2000 (after discarding 500). Global model, $\tau^2 = \{0.025, 0.00025, 0.01, 0.025, 0.025\}$; local difference, $\tau^2 = 125$; local laplacian, $\tau^2 = 45$.))

prior	β	α_0	α_1	α_2	γ_1	γ_2
global		15.01	0.1211	1.7879	24.79	23.57
local difference	0.3	15.44	0.1802	1.6036	24.61	23.60
local laplacian	5.0	15.32	0.1800	1.6096	23.73	23.29

inspection. Panel (*b*) shows part of a phantom image. The source of the hole evident in this display is a steel ball 2 cm in diameter, suspended within the perspex box (pixels in this image are approximately 0.3 cm across). In both of these panels there is a high level of noise due to the Poisson variation: the maximum counts per pixel are only 248 and 288 respectively. There is interest in the medical physics community in the development of image analysis procedures for such data that perform well even at considerably smaller photon levels than this. This naturally means a worse signal-to-noise ratio, but is attractive from the point of view of patient safety (as isotope dosage can be reduced), and to reduce motion blur due to patient movement or isotope flow. To simulate lower photon levels, we have used a thinned image,

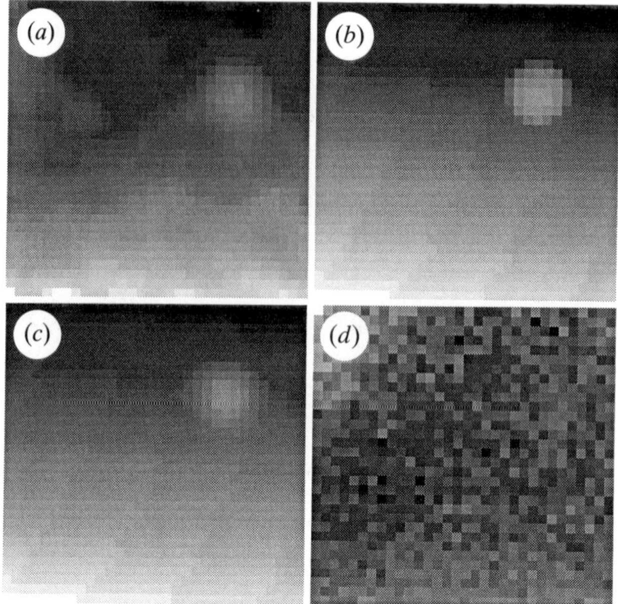

Figure 8. Reconstructions using local difference prior. (*a*) Posterior mean. (*b*) Post processed mean. (*c*) Fitted values. (*d*) Raw residuals.

displayed in figure 6*c*. This is found by binomial resampling from the image in figure 6*b*, retaining each photon, independently, with probability $\frac{1}{3}$. This clearly retains the Poisson character of the random variation, while producing a noisier image in which the maximum count per pixel is 95.

Our analysis of these data has assumed a bivariate gaussian point-spread function with $\sigma = 1.5$ pixels, truncated at $\pm 2\sigma$, a hole radius of 3.5 and depths 80 and 27 for the unthinned and thinned data-sets respectively. These values are based on the dimensions of the phantom, and information about the point-spread function provided by Dr Gibson. A similar modification to that used in our one-dimensional experiments was used to deal with edge effects, and again spatial discretization was imposed by integration, not sampling. The hyperparameter values and numbers of sweeps used are given in table 3.

Figures 7, 8 and 9 display results of our analysis of the phantom data, using each of the three methods, and figures 10, 11 and 12, the corresponding results for the binomially thinned data. In each case the estimated posterior mean is displayed, together with the post-processed output in the case of the pixel-based priors. The remaining panels display the 'fitted values' resulting from applying the point-spread function to the estimates of the truth, and the raw residuals, that is, pixelwise differences between data and fitted values. These latter images will be affected by various departures from the model: if all is well, we should expect to see random noise superimposed on a smooth trend representing deviation of the ramp from planarity. That seems to be the case in these figures. On the other hand, if the lesion had been incorrectly located, its depth or shape mis-specified, or if there had been in fact no lesions present, or more than one, then corresponding patterns would have been apparent in the residual image.

All of the methods produce similar estimates of the lesion location (γ_1, γ_2), and the ramp intercept (α_0) and gradient (α_1, α_2), on both the original and the thinned data-

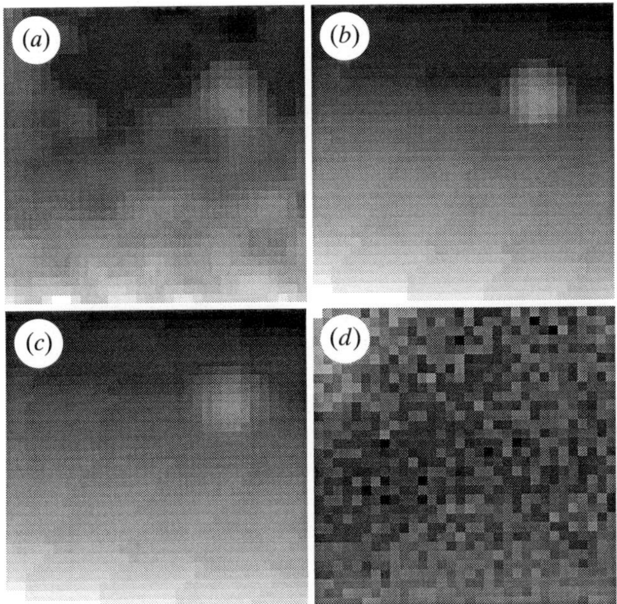

Figure 9. Reconstructions using local laplacian prior. (*a*) Posterior mean. (*b*) Post processed mean. (*c*) Fitted values. (*d*) Raw residuals.

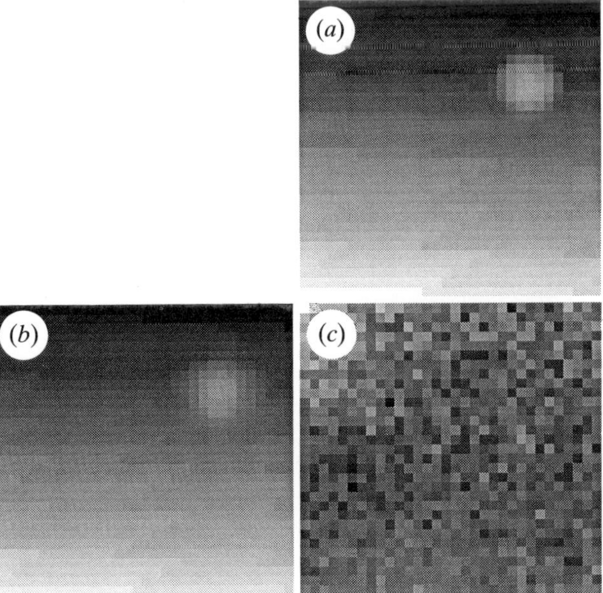

Figure 10. Reconstructions using global prior (thinned data). (*a*) Posterior mean. (*b*) Fitted values. (*c*) Raw residuals.

sets. These estimates are presented in table 3. It is particularly pleasing that even when the signal-to-noise ratio is as low as it is in the thinned data of figure 6*c*, these procedures do not break down.

In figure 13, we display traces of two monitoring statistics for each of the three

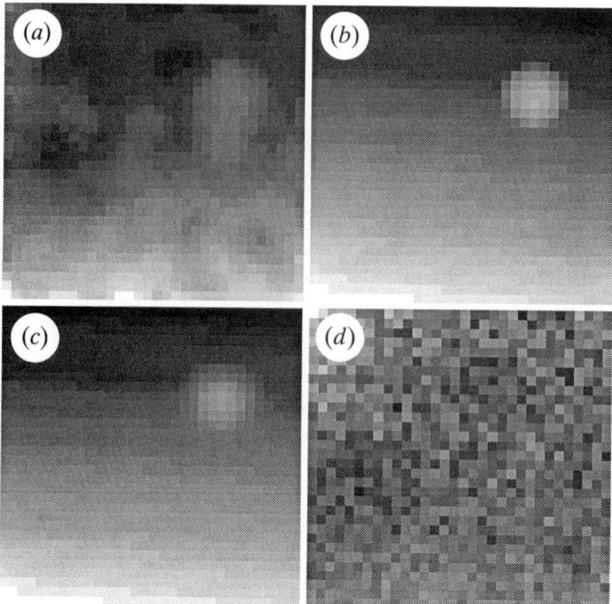

Figure 11. Reconstructions using local difference prior (thinned data). (*a*) Posterior mean. (*b*) Post processed mean. (*c*) Fitted values. (*d*) Raw residuals.

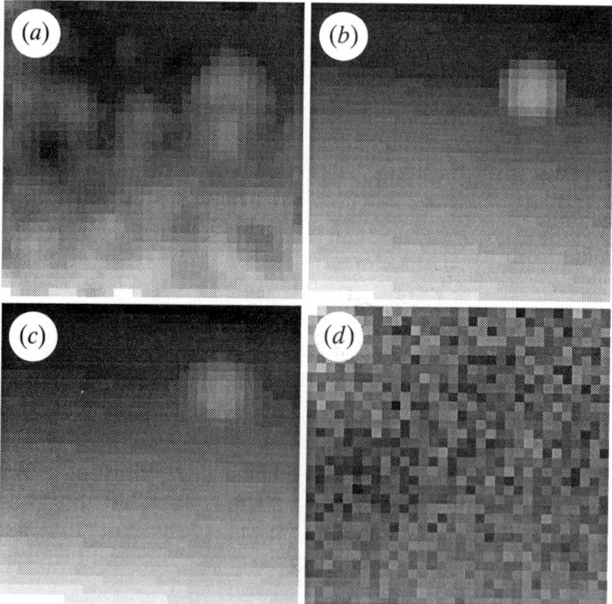

Figure 12. Reconstructions using local laplacian prior (thinned data). (*a*) Posterior mean. (*b*) Post processed mean. (*c*) Fitted values. (*d*) Raw residuals.

methods, and this time no difficulty is apparent. Monte Carlo variances for parameter estimates are given in table 4 for each of the methods.

We do not have estimated sampling variances for the estimates given in table 3, although we do intend to explore in future work a possible approach to obtaining such variances in bootstrap fashion, by exploiting the binomial resampling idea that

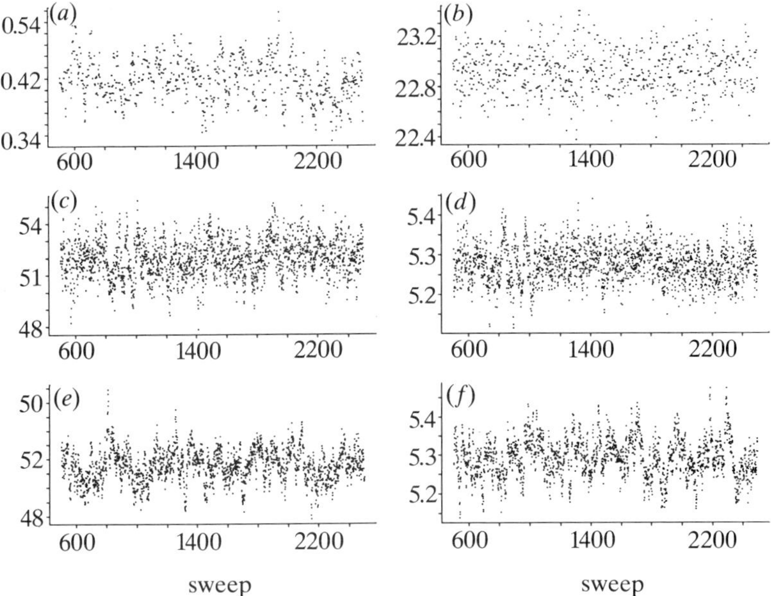

Figure 13. Monitoring statistics. Global prior: (*a*) horizontal gradient parameter; (*b*) vertical location parameter. Local difference prior: (*c*) intercept of LS plane; (*d*) vertical gradient of LS plane. Local laplacian prior: (*e*) intercept of LS plane; (*f*) vertical gradient of LS plane.

Table 4. *Monte Carlo variances*

Phantom data

prior	k	b	α_0	α_1	α_2	γ_1	γ_2
global	50	40	0.00903	1.229×10^{-5}	1.671×10^{-5}	6.316×10^{-5}	4.634×10^{-5}
local difference	100	20	0.00542	0.781×10^{-5}	0.816×10^{-5}	0.004432	0.0040727
local laplacian	50	40	0.01217	2.563×10^{-5}	2.330×10^{-5}	0.045221	0.0194146

Thinned phantom data

prior	k	b	α_0	α_1	α_2	γ_1	γ_2
global	50	40	1.513×10^{-3}	2.436×10^{-6}	4.012×10^{-6}	9.557×10^{-5}	2.797×10^{-4}
local difference	50	40	0.592×10^{-3}	1.521×10^{-6}	1.521×10^{-6}	0.19339	0.098933
local laplacian	100	20	4.146×10^{-3}	6.439×10^{-6}	5.681×10^{-6}	0.37729	0.119511

was used to produce the thinned data-set. Estimated sampling variances for maximum likelihood estimation of $(\alpha_0, \alpha_1, \alpha_2)$ in fitting the planar ramp, without the hole, to the unthinned data of figure 6*b* are approximately (0.724, 0.00142, 0.00147) respectively.

We are indebted to Chris Gibson for providing the data, and to the referees for their help in improving the paper. We acknowledge financial support for R.G.A. and provision of computing equipment by the Science and Engineering Research Council, through its Complex Stochastic Systems initiative.

References

Amit, Y., Grenander, U. & Piccioni, M. 1991 Structured image restoration through deformable templates. *J. Am. statist. Ass.* (In the press.)

Besag, J. 1974 Spatial interaction and the statistical analysis of lattice systems (with discussion). *Jl R. statist. Soc.* B **36**, 192–236.

Besag, J. 1983 Discussion of paper by P. Switzer. *Bull. Int. statist. Inst.* **50** (3), 422–425.

Besag, J. 1986 On the statistical analysis of dirty pictures (with discussion). *Jl R. statist. Soc.* B **48**, 259–302.

Besag, J., York, J. & Mollié, A. 1991 Bayesian image restoration, with two applications in spatial statistics (with discussion). *Ann. Inst. statist. Math.* **43**, 1–59.

Chow, Y., Grenander, U. & Keenan, D. M. 1988 Hands: a pattern theoretic study of biological shape. *Research notes in neural computing*, vol. 2. Berlin: Springer.

Clifford, P. 1986 Contribution to discussion of paper by J. Besag. *Jl R. statist. Soc.* B **48**, 284.

Dupuis, P., Geman, D., Horowitz, J. & Reynolds, G. 1991 Statistical inference on the shape of circumstellar disks from HST observations. Working paper, University of Massachusetts, Amherst, U.S.A.

Gelfand, A. E. & Smith, A. F. M. 1990 Sampling-based approaches to calculating marginal densities. *J. Am. statist. Ass.* **85**, 398–409.

Geman, D. 1991 Random fields and inverse problems in imaging. *Lecture Notes in Math.* Berlin: Springer. (In the press.)

Geman, S. & Geman, D. 1984 Stochastic relaxation, Gibbs distributions and the Bayesian restoration of images. *IEEE Trans. Pattern Analys. Mach. Intell.* **6**, 721–741.

Geman, S. & McClure, D. 1987 Statistical methods for tomographic image reconstruction. *Bull. Int. statist. Inst.* **52** (4), 5–21.

Green, P. J. 1986 Contribution to discussion of paper by J. Besag. *Jl R. statist. Soc.* B **48**, 284–285.

Green, P. J. 1987 Penalized likelihood for general semi-parametric regression models. *Int. Statist. Rev.* **55**, 245–259.

Green, P. J. 1990 Bayesian reconstructions from emission tomography data using a modified EM algorithm. *IEEE Trans. Medical Imaging* **9**, 84–93.

Green, P. J. & Han, X.-L. 1991 Metropolis methods, gaussian proposals and antithetic variables. Stochastic models, statistical methods and algorithms in image analysis (ed. P. Barone, A. Frigessi & M. Piccioni). *Lecture Notes in Statist.* Berlin: Springer. (In the press.)

Hammersley, J. M. & Handscomb, D. C. 1964 *Monte Carlo methods.* London: Methuen.

Hastings, W. K. 1970 Monte Carlo simulation methods using Markov chains, and their applications. *Biometrika* **57**, 97–109.

Jennison, C. & Jubb, M. 1991 Aggregation and refinement in binary image restoration. In *Spatial statistics and imaging* (ed. A. Possolo), IMS Lecture Notes. (In the press.)

Jones, M. C. & Silverman, B. W. 1989 An orthogonal series density estimation approach to reconstructing position emission tomography images. *J. appl. Statist.* **16**, 177–191.

Larsson, S. A. 1980 Gamma camera emission tomography. *Acta radiologica Suppl.* 363.

Qian, W. & Titterington, D. M. 1991*a* Stochastic relaxations and E-M algorithms for Markov random fields. *J. statist. Comput. Simulation.* (In the press.)

Qian, W. & Titterington, D. M. 1991*b* Estimation of parameters in hidden Markov models. *Phil. Trans. R. Soc. Lond.* A **337**, 407–428. (This volume.)

Ripley, B. D. & Sutherland, A. I. 1990 Finding spiral structures in images of galaxies. *Phil. Trans. R. Soc. Lond.* A **332**, 477–485.

Sheehan, N. A. & Thomas, A. W. 1991 On the irreducibility of a Markov chain defined on a space of genotype configurations by a sampling scheme. (Submitted.)

Shepp, L. A. & Vardi, Y. 1982 Maximum likelihood reconstruction for emission tomography. *IEEE Trans. Medical Imaging* **1**, 113–122.

Smith, A. F. M. 1991 Bayesian computational methods. *Phil. Trans. R. Soc. Lond.* A **337**, 369–386. (This volume.)

Sokal, A. D. 1989 Monte Carlo methods in statistical mechanics: foundations and new algorithms. Cours de Troisième Cycle de la Physique en Suisse Romande, Lausanne.

Network routing

By F. P. Kelly
Statistical Laboratory, University of Cambridge, 16 Mill Lane, Cambridge CB2 1SB, U.K.

How should flows through a network be organized, so that the network responds sensibly to failures and overloads? The question is currently of considerable technological importance in connection with the development of computer and telecommunication networks, while in various other forms it has a long history in the fields of physics and economics. In all of these areas there is interest in how simple, local rules, often involving random actions, can produce coherent and purposeful behaviour at the macroscopic level. This paper describes some examples from these various fields, and indicates how analogies with fundamental concepts such as energy and price can provide powerful insights into the design of routing schemes for communication networks.

1. Introduction

Modern computer and telecommunication networks are able to respond to randomly fluctuating demands and failures by rerouting traffic and by reallocating resources. They are able to do this so well that, in many respects, large-scale networks appear as coherent, almost intelligent, organisms. The design and control of such networks require an understanding of a variety of fundamental issues, and this is providing an important stimulus to many areas of mathematics and engineering.

To give an example of current importance, a major practical and theoretical issue concerns the extent to which control can be decentralized. Over a period of time the form of the network or the demands placed on it may change, and routings may need to respond accordingly. It is rarely the case, however, that there should be a central decision-making processor, deciding upon these responses. Such a centralized processor, even if it were itself completely reliable and could cope with the complexity of the computational task involved, would have its lines of communication through the network vulnerable to delays and failures. Rather, decision-making should be decentralized and of a simple form: the challenge is to understand how such decentralized decision-making can be organized so that the network as a whole reacts intelligently to outside stimuli.

The behaviour of large-scale systems has been of great interest to mathematicians for over a century, with many examples coming from physics. The behaviour of a gas can be described at the microscopic level in terms of the position and velocity of each molecule. At this level of detail a molecule's velocity appears as a random process, with a stationary distribution as found by Maxwell (and later discussed by Erlang (1925)). Consistent with this detailed microscopic description of the system is macroscopic behaviour best described by quantities such as temperature and pressure. Similarly the behaviour of electrons in an electrical network can be described in terms of random walks, and yet this simple description at the

Phil. Trans. R. Soc. Lond. A (1991) **337**, 343–367
Printed in Great Britain

microscopic level leads to rather sophisticated behaviour at the macroscopic level: the pattern of potentials in a network of resistors is just such that it minimizes heat dissipation for a given level of current flow (Thomson & Tait 1879). The local, random behaviour of the electrons causes the network as a whole to solve a rather complex optimization problem.

Of course simple local rules may lead to poor system behaviour if the rules are the wrong ones. Road traffic networks provide a chastening example of this. Braess's paradox describes how, if a new road is added to a congested network, the average speed of traffic may fall rather than rise, and indeed everyone's journey time may lengthen. The paradox may actually have occurred during 'development' in the centre of Stuttgart (Knödel 1969), and counterintuitive consequences of road closures are often reported (*New York Times* 1990). The attempts of individuals to do the best for themselves lead to everyone suffering. It is possible to alter the local rules, by the imposition of appropriate tolls, so that the network behaves more sensibly, and indeed road traffic networks have long provided a key example of the economic principle that externalities need to be appropriately penalized if the invisible hand is to lead to optimal behaviour (Pigou 1920; Walters 1961).

A telephone network provides a fascinating example of a large-scale system where strange effects can occur. For instance, suppose that 'intelligent' exchanges react to blocked routes by rerouting calls along more resource-intensive paths. This in turn may cause later calls to be rerouted, and the cascade effect may lead to a catastrophic change in the network's behaviour. When exchanges strive to be efficient there is a possibility they may overdo it. In some respects the network's behaviour resembles water boiling. Just as a small change in the temperature of a body of water can cause a pronounced macroscopic effect, so a small change in the load on a network can produce an unexpected and massive failure. This discussion indicates the care that must be taken with the development of routing rules for large networks.

There is currently considerable interest in the similarities between complex systems from diverse areas of physics, economics and biology, and it is clear that topics such as noisy optimization and adaptive learning provide mathematical metaphors of value across many fields. The reader is referred to Pines (1987), Anderson *et al.* (1988) and Langton (1989) for the lively and thought-provoking proceedings of a series of workshops, and to Whittle (1986) for a study of statistical equilibrium in systems of interacting components that is both broad and penetrating. Our aim in this paper is, by comparison, much narrower: to describe within a common framework some examples from three areas, and thus to indicate how earlier, often much earlier, work on electrical networks and traffic flow has influenced recent work on routing schemes for communication networks.

In §2 we outline the connections among random walks, electrical networks and variational principles. We describe how an interacting particle system, precisely defined at a microscopic level in terms of simple, randomized, local rules, can also be described at an intermediate level of aggregation in terms of Ohm's Law and Kirchhoff's equations, and at a macroscopic level in terms of energy minimization. The interacting particle system we describe is due to Kingman (1969), and our treatment of the associated optimization problem derives from Whittle (1971). For a beautifully written account of the area, and of its history back to Lords Rayleigh and Kelvin, the reader is referred to Doyle & Snell (1984).

In §3 we consider multiclass flow models, including queueing and road traffic networks. If customers or drivers can choose their routes, then exercise of this choice may force the system to a competitive equilibrium. If drivers attempt to minimize their own delay the resulting equilibrium flows will minimize a certain objective function defined for the network. However, the objective function is certainly not the average network delay: this is dramatically illustrated by Braess's (1968) paradox, outlined above. We describe a variant of this paradox: if drivers are provided with extra information about random delays ahead, the outcome may well be a new equilibrium in which delays are increased for everyone. Traffic-dependent tolls are sufficient to force the system to an equilibrium which minimizes average network delay: the tolls charge drivers for the delays they cause to others. The study of appropriate tolls has long been a topic of central importance in economics, and it is interesting to note that Pigou (1920, p. 194) used a simple two-node, two-link traffic network to illustrate the possibility that taxation could 'create an "artificial" situation superior to the "natural" one'. (We note in passing that developments in electronics now make feasible the practical application of both route guidance and road pricing (van Vuren & Smart 1990).) Potts & Oliver (1972) survey the characterization of equilibrium flow patterns as extremal values, and Nagurney (1987) provides a recent review of competitive equilibrium problems including more general traffic network models and related models of economic markets.

In the remainder of the paper we outline more recent work on the modelling of telecommunications networks. In §4 we show how the microscopic description of a telephone network, in terms of random arrival streams and rules for accepting and routing calls, leads the network to behave as if it is attempting to minimize a certain potential function. However, just as in road traffic networks, the potential function implicitly minimized may bear little relation to the network performance criteria of interest to system designers. Indeed Wroe *et al.* (1990) have described an example very similar to Braess's paradox, that has arisen in the design of the BT international access network, where the addition of capacity to a network causes the performance to become worse. Various other forms of perverse behaviour can be interpreted in terms of an implicit minimization: for example, in situations where alternative routing causes the potential function to have multiple minima, a slight change in traffic load may cause the network to jump catastrophically between minima. In §5 we discuss how explicit consideration of network performance criteria can lead, through notions of shadow prices and implied costs, to decentralized adaptive routing schemes which are at least attempting to optimize the right function.

Might it be possible to choose the microscopic rules governing the behaviour of a telephone network, the rules for accepting and routing calls, so that the network is implicitly optimizing sensible performance criteria, much as current flow in an electrical network is implicitly minimizing heat dissipation? This is a difficult and wide-ranging question, but at least in some circumstances it can be answered positively. In §6 we describe a scheme which has been developed for the British telephone network by researchers at Cambridge and at British Telecom's laboratories at Martlesham. The scheme, known as Dynamic Alternative Routing (DAR), is now being implemented in the British trunk network (Stacey & Songhurst 1987; Gibbens 1988; Gibbens *et al.* 1988; Key & Whitehead 1988). The scheme will lessen the impact of forecasting errors, make better use of spare capacity and respond robustly to failures and overloads. It will also permit flexible use of network resources enabling, for example, the rapid introduction of new services where demand is often uncertain. DAR uses very simple rules across the network, making constructive use of inherent randomness to search out good routing patterns. In this way the network itself operates as a distributed computer, executing a highly parallel randomized algorithm to solve a complex optimization problem.

2. Random walk and electrical networks

We begin this section by describing a simple flow model due to Kingman (1969) and further discussed by Kelly (1979). The model can be viewed as a naive description of the movement of electrons in a conductor, and we shall phrase our discussion in terms of familiar electrical concepts such as current and potential. We use the model to develop the connections between random walks, electrical networks and variational principles.

Consider, then, the following interacting particle system. There is a set of sites, J, and each site $j \in J$ may be empty or may be occupied by a single particle. If site j is occupied and site k is empty then, with probability intensity λ_{jk} $(= \lambda_{kj})$, the particle at site j moves to site k. If site j is occupied then, with probability intensity μ_j, the particle at site j leaves the system entirely. If site k is empty then, with probability intensity ν_k, a particle arrives at site k from outside the system. These rules define a finite state Markov process, about which we can ask a number of questions. What is the stationary probability p_j that site j is occupied? What is the average net rate of flow of particles from site j to site k?

To answer these questions consider the following button model. Append to the set of sites J two further sites, labelled 0 and 1, and let each site contain a single button. The buttons are distinguishable; we can imagine them to be of different colours. The buttons occupying sites j and k interchange positions with probability intensity λ_{jk}, for $j, k \in J \cup \{0, 1\}$, where

$$\lambda_{0j} = \lambda_{j0} = \mu_j, \quad \lambda_{1j} = \lambda_{j1} = \nu_j, \quad j \in J,$$

and $\lambda_{01} = \lambda_{10} = 0$. We see that any particular button performs a symmetric (and hence reversible) random walk around the sites of the system. Now imagine that buttons entering site 1 are painted black, while buttons entering site 0 are painted white. If A is the set of those sites $j \in J$ which contain a black button then A behaves stochastically just as does the set of occupied sites in the earlier interacting particle system. Thus to find the probability p_j that site j is occupied we need only look backwards through time at the earlier movements of the button which currently occupies site j. These movements form a symmetric random walk starting from site j with transition intensities λ_{jk} for $j, k \in J \cup \{0, 1\}$; p_j is equal to the probability that this random walk reaches site 1 before site 0. Considering where the first step of the random walk takes the button leads to the equations

$$p_j = \sum_k \frac{\lambda_{jk}}{\sum_i \lambda_{ji}} p_k, \quad j \in J,$$

$$p_0 = 0, \quad p_1 = 1.$$

We can rewrite these equations as

$$\sum_k \lambda_{jk}(p_j - p_k) = 0, \quad j \in J, \tag{2.1}$$

$$p_0 = 0, \quad p_1 = 1. \tag{2.2}$$

Equations (2.1) and (2.2), however, are precisely Kirchhoff's equations for an electrical network with nodes from the set $J \cup \{0, 1\}$: just interpret p_j as the electrical potential of node j, connect nodes j and k by a wire of resistance λ_{jk}^{-1}, and hold nodes 0 and 1 at potentials 0 and 1 respectively. Equation (2.1) simply states that the total current flowing out of node j is zero.

The average net flow of particles from site j to site k is

$$\lambda_{jk} P\{\text{site } j \text{ occupied and site } k \text{ empty}\} - \lambda_{kj} P\{\text{site } j \text{ empty and site } k \text{ occupied}\}$$
$$= \lambda_{jk}(p_j - p_k).$$

Put more formally, the system is a finite state Markov chain, and hence the net flow of particles from site j to site k over a time interval $[0, T]$ will almost surely converge as $T \to \infty$ to

$$u_{jk} = \lambda_{jk}(p_j - p_k). \tag{2.3}$$

This, however, is precisely the current flowing from node j to node k in the electrical network; equation (2.3) is just Ohm's law. The average flow of particles through the network can be calculated from either the flow out of node 1, or the flow into node 0, and is

$$U = \sum_k \lambda_{1k}(1 - p_k) = \sum_j \lambda_{j0}\, p_j, \tag{2.4}$$

which is precisely the total current flowing through the electrical network.

The energy dissipated by a potential difference of $p_j - p_k$ across a wire of resistance λ_{jk}^{-1} is the product of the current and voltage, and is thus $\lambda_{jk}(p_j - p_k)^2$. The energy dissipation in the above electrical network is then

$$\tfrac{1}{2} \sum_{j,k} \lambda_{jk}(p_j - p_k)^2,$$

where the summation runs over $j, k \in J \cup \{0, 1\}$, and the factor $\frac{1}{2}$ is necessary since each edge is counted twice in the sum. Consider the following problem.

$$\text{Minimize} \quad \tfrac{1}{2} \sum_{j,k} \lambda_{jk}(p_j - p_k)^2, \tag{2.5a}$$

$$\text{over} \quad (p_j, j \in J), \tag{2.5b}$$

$$\text{subject to} \quad p_0 = 0, \quad p_1 = 1. \tag{2.5c}$$

The objective function (2.5*a*) is strictly convex, and a differentiation yields that the unique optimum is given by the solution to Kirchhoff's equations (2.1) and (2.2). This is the *Dirichlet principle*: the potentials taken within the electrical network minimize the total energy dissipation.

The energy dissipated by a current flow u_{jk} through a wire of resistance λ_{jk}^{-1} is u_{jk}^2/λ_{jk}. Consider, then, the following problem, where the objective function is again the energy dissipation of the network, but now expressed in terms of currents rather than potentials.

$$\text{Minimize} \quad \tfrac{1}{2} \sum_{j,k} u_{jk}^2/\lambda_{jk}, \tag{2.6a}$$

$$\text{over} \quad u_{jk}(= -u_{kj}), \quad j, k \in J \cup \{0, 1\}, \tag{2.6b}$$

$$\text{subject to} \quad \sum_k u_{jk} = \begin{cases} 0 & j \in J, \\ -U & j = 0, \\ U & j = 1. \end{cases} \tag{2.6c}$$

The conditions (2.6*c*) require a flow U of current into the network from node 1, a flow U of current out of the network to node 0, and that flow be balanced at nodes $j \in J$. Again the objective function is strictly convex: hence the optimum is unique and can be obtained by differentiating the *lagrangian form*

$$L(u;p) = \tfrac{1}{2}\sum_{j,k} u_{jk}^2/\lambda_{jk} - 2\sum_{j} p_j\left(\sum_{k} u_{jk}\right),$$

where $p_j, j \in J \cup \{0, 1\}$, are now Lagrange multipliers, and where for later convenience we have introduced an extra factor of 2 before the constraints.

Differentiating L with respect to u_{jk} we obtain

$$\partial L/\partial u_{jk} = 2(u_{jk}/\lambda_{jk} - p_j + p_k),$$

since $u_{kj} = -u_{jk}$. Thus the optimum takes the form

$$u_{jk} = \lambda_{jk}(p_j - p_k), \quad j, k \in J \cup \{0, 1\},$$

for a choice of Lagrange multipliers p_j that cause the resulting u_{jk} to satisfy the constraints (2.6*c*); but the p_js that solve (2.1) and (2.2) *are* such a choice. Thus the Lagrange multipliers should be set equal to the electrical potentials, and the resulting flows u_{jk} are optimal. This is *Thomson's principle*: the flow pattern of current within an electrical network is that which minimizes the energy dissipation over all flow patterns achieving the same total current.

Together the Dirichlet principle and Thomson's principle provide one of the many examples from physics of complementary variational principles (Whittle 1971). From the viewpoint of optimization theory, if the problem (2.6) is recast as one of maximizing flow for a given energy dissipation then it is possible to obtain the problem (2.5) as the formal lagrangian dual.

In this section we have seen an example of a system which can be viewed at various levels of abstraction. At the microscopic level, it can be viewed in terms of particles performing simple random walks. If average rates of particle flow are studied, then these flows are given by Kirchhoff's equations (2.1), (2.2), and Ohm's law (2.3). Finally, at the network level, the flow patterns that emerge solve a constrained optimization problem whose objective function is the network's energy dissipation.

Thomson's principle, in particular, is an extremely useful and suggestive result. It follows directly that if a link is added to an electrical network and the same total current is carried then the energy dissipation is reduced: after all, the old flow pattern remains feasible for the new optimization problem. Might it be possible to design telecommunication networks similarly, so that, for example, additions of capacity are necessarily helpful? Before considering this question further, we look first at queueing networks, and the ways in which implicit optimization can go wrong.

3. Queueing networks and traffic flow

In the interacting particle system of the last section the particles behaved similarly, each attempting to perform a symmetric random walk. In this section we study a more complicated system, where particles are of different classes and where class determines the direction of flow through the network.

We begin with a brief description of an open multiclass network of $\cdot/M/1$ queues (for a more leisurely introduction see Kelly (1979)). Let the network have a set J of queues, and suppose that a customer entering the system is labelled with the route he will follow through the network. More specifically, suppose that customers labelled $r = (r(1), r(2), \dots, r(M_r))$ arrive at the system in a Poisson stream of rate ν_r and pass through the sequence of queues

$$r(1), r(2), \dots, r(M_r),$$

before leaving the system. Thus the queue which a customer labelled r visits at stage $m\,(=1, 2, \dots, M_r)$ of his passage through the network is queue $r(m)$. Write R for the set of possible routes, and assume that as r varies over R it indexes independent Poisson arrival streams. Let each queue have a single server operating a first come first served discipline, and suppose that a customer visiting queue j has a service time there which is exponentially distributed with parameter ϕ_j and independent of all other service times and of the arrival streams at the network.

Write $j \in r$ if route r passes through queue j at some stage, and, for simplicity of notation, suppose no route passes through the same queue more than once. Let

$$\rho_j = \sum_{r:j\in r} \nu_r,$$

and suppose $\rho_j < \phi_j, j = 1, 2, \dots, J$. Thus ρ_j measures the throughput of queue j.

Let $n_j(t)$ be the number of customers in queue j at time t, and let $n(t) = (n_j(t), j\in J)$. Then the (non-Markov) stochastic process $(n(t), t \geqslant 0)$ has a unique stationary distribution, and under this distribution $\pi(n) = P\{n(t) = n\}$ is given by

$$\pi(n) = \prod_{j\in J} \pi_j(n_j), \tag{3.1}$$

where
$$\pi_j(n_j) = (1-\rho_j/\phi_j)\,(\rho_j/\phi_j)^{n_j}, \tag{3.2}$$

the geometric distribution familiar as the stationary distribution of an $M/M/1$ queue with arrival rate ρ_j. From the distribution (3.2) and Little's formula it follows that the mean sojourn time of a customer in queue j is

$$D_j(\rho_j) = (\phi_j - \rho_j)^{-1}. \tag{3.3}$$

Sometimes we prefer to observe the network from the point of view of customers labelled r. When a typical customer labelled r arrives at a queue j on his route, the probability he finds n_j customers already in that queue is given by expression (3.2); the sojourn time of a typical customer labelled r in a queue j on his route is exponentially distributed with parameter $\phi_j - \rho_j$.

A wide range of more general queueing networks share properties of the simple network described above, that the stationary distribution for the numbers in different queues has the product-form (3.1), and that the mean sojourn time of a customer in queue j is a function $D_j(\rho_j)$ of the throughput of queue j.

Suppose now that it is possible to vary the parameters ν_r. For example, if the model represents a packet-switched communication network, there may be a variety of possible routes r through the network capable of linking two nodes, and the network may be able to shift traffic to routes with lower mean delays. More formally, let the label s of a customer arriving at the network identify not a single route, but a set of routes, any of which could serve the customer. We can view s as labelling a source-sink node pair. Set $H_{sr} = 1$ if $r \in s$, so that a customer labelled s can be served by the route r, and set $H_{sr} = 0$ otherwise. This defines a $0-1$ matrix $H = (H_{sr}, s\in S,$

$r \in R$). For each $r \in R$ let $s(r)$ identify the unique value $s \in S$ such that $H_{sr} = 1$; we thus view $s(r)$ as the source-sink node pair served by route r. Let $A_{jr} = 1$ if route r passes through link j, and set $A_{jr} = 0$ otherwise. This defines a $0-1$ matrix $A = (A_{jr}, j \in J, r \in R)$.

A *Wardrop equilibrium* is a collection $\nu = (\nu_r, r \in R), \rho = (\rho_j, j \in J)$ of non-negative numbers such that

$$H\nu = b, \tag{3.4a}$$

$$A\nu = \rho, \tag{3.4b}$$

and

$$\nu_r > 0 \Rightarrow \sum_{j \in r} D_j(\rho_j) = \min_{r' \in s(r)} \sum_{j \in r'} D_j(\rho_j), \quad r \in R. \tag{3.4c}$$

Equation (3.4*a*) states that the traffic over routes r serving the node pair s sums to b_s, while equation (3.4*b*) states that the traffic over routes through link j sums to ρ_j. The implication (3.4*c*) expresses the defining characteristic of a Wardrop equilibrium (Wardrop 1952), that if a route r is actively used, it achieves the minimum delay over all routes serving the node pair $s(r)$.

Does a Wardrop equilibrium exist, and, if so, is it unique ? To answer this question, consider the following optimization problem.

$$\text{Minimize} \quad \sum_{j \in J} \int_0^{\rho_j} D_j(z)\, \mathrm{d}z, \tag{3.5a}$$

$$\text{over} \quad \nu, \rho \geqslant 0, \tag{3.5b}$$

$$\text{subject to} \quad H\nu = b, \quad A\nu = \rho. \tag{3.5c}$$

Impose the mild condition that $D_j(z)$ is continuous and strictly increasing. Then the optimum can be found by differentiating the lagrangian form

$$L(\nu, \rho; \lambda, \mu) = \sum_{j \in J} \int_0^{\rho_j} D_j(z)\, \mathrm{d}z + \lambda^{\mathrm{T}}(b - H\nu) - \mu^{\mathrm{T}}(\rho - A\nu),$$

where λ, μ are vectors of Lagrange multipliers. But

$$\frac{\partial L}{\partial \nu_r} = -\lambda_{s(r)} + \sum_{j \in r} \mu_j,$$

and

$$\partial L / \partial \rho_j = D_j(\rho_j) - \mu_j.$$

Hence a maximum of L over $\nu, \rho \geqslant 0$ occurs when

$$\mu_j = D_j(\rho_j)$$

and

$$\lambda_{s(r)} = \sum_{j \in r} \mu_j \quad \text{if} \quad \nu_r > 0$$

$$\leqslant \sum_{j \in r} \mu_j \quad \text{if} \quad \nu_r = 0.$$

The Lagrange multipliers have a simple interpretation: μ_j is the delay on link j, and λ_s is the minimum delay over all routes serving the node pair s. The minima of the objective function (3.5*a*) correspond precisely to Wardrop equilibria. Since $D_j(z)$ is strictly increasing the objective function (3.5*a*) is a strictly convex function of ρ.

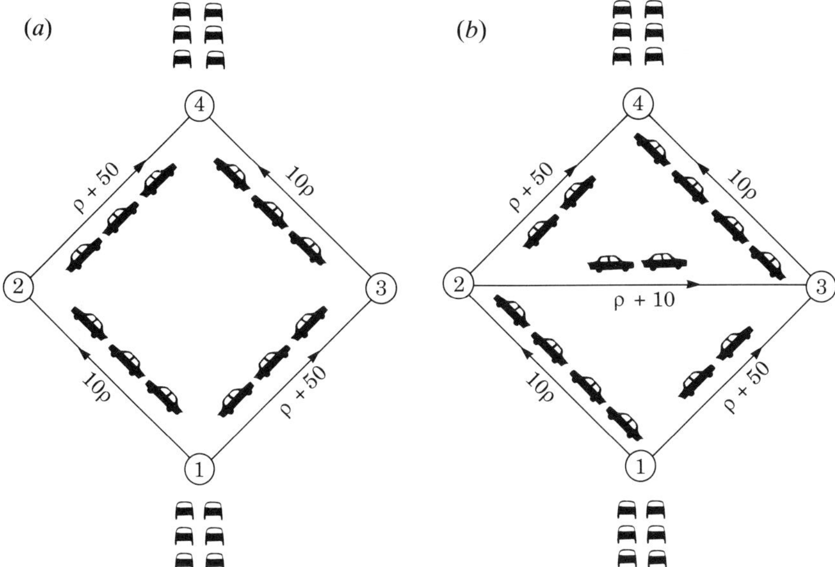

Figure 1. Braess's paradox. The addition of a link causes everyone's journey time to lengthen. (After Braess 1968; Cohen 1988; Cohen & kelly 1990.)

Hence there is a unique optimum for the flow vector ρ, although there may be many corresponding values of the non-negative vector ν satisfying the linear relations (3.5c).

Thus, if traffic in the network distributes itself towards routes with lower mean delays, the equilibrium flows $\rho = (\rho_j, j \in J)$ will solve the optimization problem (3.5). This does *not* mean that average delays in the network will be minimal: and a striking illustration of this fact is provided by Braess's paradox (Braess 1968; Cohen & Kelly 1990). Consider the network illustrated in figure 1a. Cars travel from node 1 to node 4, via either node 2 or node 3. The total flow is 6, and the link delays $D_j(\rho)$ are given next to links in the figure. The Wardrop equilibrium is shown. Now suppose that a new link is added, between nodes 2 and 3, as shown in figure 1b. Traffic is attracted onto the new link, and the new Wardrop equilibrium is shown in figure 1b. Observe that each car incurs a delay of 83 in figure 1a, while each care incurs a delay of 92 in figure 1b. Adding the new link has increased everyone's delay!

The point, of course, is that, while the Wardrop equilibrium minimizes the function (3.5a), this objective function is not closely related to average delays. Consider next the following problem.

$$\text{Minimize} \quad \sum_{j\in J} \rho_j D_j(\rho_j), \tag{3.6a}$$

$$\text{over} \quad \nu, \rho \geqslant 0, \tag{3.6b}$$

$$\text{subject to} \quad H\nu = b, \quad A\nu = \rho. \tag{3.6c}$$

Note that the objective function (3.6a) now measures average network delay. Impose the fairly mild condition that the function (3.6a) is convex and differentiable. The lagrangian for this problem is

$$L(\nu, \rho; \lambda, \mu) = \sum_{j\in J} \rho_j D_j(\rho_j) + \lambda^{\mathrm{T}}(b - H\nu) - \mu^{\mathrm{T}}(\rho - A\nu).$$

Again

$$\frac{\partial L}{\partial \nu_r} = -\lambda_{s(r)} + \sum_{j \in r} \mu_j,$$

but now

$$\partial L / \partial \rho_j = D_j(\rho_j) + \rho_j D_j'(\rho_j) - \mu_j.$$

Hence a maximum of L over $\nu, \rho \geqslant 0$ occurs when

$$\mu_j = D_j(\rho_j) + \rho_j D_j'(\rho_j),$$

and

$$\begin{aligned} \lambda_{s(r)} &= \sum_{j \in r} \mu_j \quad \text{if} \quad \nu_r > 0 \\ &\leqslant \sum_{j \in r} \mu_j \quad \text{if} \quad \nu_r = 0. \end{aligned}$$

Again the Lagrange multipliers have a simple interpretation. Suppose that, in addition to the delay $D_j(\rho_j)$, users of link j incur a traffic-dependent toll

$$T_j(\rho_j) = \rho_j D_j'(\rho_j). \tag{3.7}$$

Then μ_j is the combined cost of using link j, and λ_s is the minimum cost over all routes serving the node pair s. If users select routes in an attempt to minimize the sum of their tolls and their delays, then they will produce a flow pattern which minimizes the average delay in the network (3.6*a*).

In an electrical network, the addition of an extra wire connecting two nodes cannot increase the energy dissipation associated with a given current flow. We have seen that this follows from Thomson's principle, since the flow pattern before addition of the wire remains feasible afterwards. Similarly, in a traffic network with tolls (3.7) the addition of a new link cannot increase average network delay, since the old flow pattern remains feasible for the new optimization problem (3.6), and so if the new flow pattern is different it must improve the objective function (3.6*a*).

In our above discussion of equilibrium flow patterns we have assumed that users are aware of the average delays $D_j(\rho_j)$, and perhaps tolls $T_j(\rho_j)$, along different routes. What if, instead, users do not know these quantities precisely, but have to rely on their previous experience along routes? To provide a clear framework for our discussion of this question, consider again the multiclass queueing network described at the beginning of this section. Suppose that a user travelling between a source–sink pair s makes this journey repeatedly, but at times separated by long intervals, and that the user observes queue lengths or sojourn times at queues as he passes through. From these observations a user can estimate average delays $D_j(\rho_j)$ at queues $j \in r$, for each route $r \in s$. If users attempt to minimize their own delays across the network then it is reasonable to expect the stationary behaviour of the network to be described by the multiclass queueing network described earlier, with stationary distribution (3.1)–(3.2) where $\rho = (\rho_j, j \in J)$ is the Wardrop equilibrium. This type of equilibrium is often termed *adaptive* or *quasi-static*: on a short timescale arrivals at the network are adequately described by Poisson streams of rates $\nu_r, r \in R$, while the rates themselves adapt over a longer timescale.

Next consider the case where tolls are imposed to encourage traffic towards a system optimal flow pattern. One possibility is that queue j could itself estimate ρ_j, the traffic through it, and hence estimate the correct toll $T_j(\rho_j)$. Such an approach has been carefully investigated in the important paper of Gallager (1977), where an implementation particularly suited to data networks is developed, and adaptive convergence to the system optimal flow pattern is established. Another possibility is that queue j might charge a toll $t_j(n)$ dependent on the number n in queue j just after a user has arrived, or a toll $t_j(D)$ dependent on the actual sojourn D of the user in queue j. In either case it will be required that the expected toll be $T_j(\rho_j)$, in order that the system optimal flow pattern be encouraged. Now $n-1$ has the geometric distribution (3.1), and D has an exponential distribution with parameter $(\phi_j-\rho_j)$. Hence we can deduce that the linear tolls

$$t_j(n) = \rho_j n/\phi_j(\phi_j-\rho_j), \quad n = 1, 2 \ldots, \tag{3.8a}$$

$$t_j(D) = \rho_j D/(\phi_j-\rho_j), \quad D \geqslant 0, \tag{3.8b}$$

lead to the correct expected tolls. These linear tolls also have the property that they charge a user precisely the externality he causes: for example, the toll (3.8a) can be shown to be the mean additional delay caused to other users of queue j if an additional user is added to queue j bringing its queue size to n. Note that linear tolls may be easy to enforce: for example, the toll (3.8b) corresponds to charging each user a flat rate $\rho_j/(\phi_j-\rho_j)$ per unit time he spends in queue j. Against this it can be argued that the tolls depend on ρ_j and ϕ_j; while the queue may know its service rate ϕ_j, if it is forced to estimate ρ_j then a statistically more efficient feedback signal to users would be the average toll $T_j(\rho_j)$.

Do there exist tolls which do not require queue j to estimate mean levels of traffic? More formally, are there tolls $t_j(n)$ or $t_j(D)$, possibly dependent on ϕ_j, which have expectation $T_j(\rho_j)$ for all values of ρ_j, but which do not depend on ρ_j? In fact

$$t_j(n) = n(n-1)/2\phi_j, \quad n = 1, 2 \ldots, \tag{3.9a}$$

$$t_j(D) = \tfrac{1}{2}\phi_j D^2 - D, \quad D \geqslant 0, \tag{3.9b}$$

are the unique functions of n or D, respectively, which have expectation $T_j(\rho_j)$ for every $\rho_j > 0$. That they have the correct expectations follows by calculation from the geometric and exponential distribution of $n-1$ and D, respectively; that they are unique follows since the geometric and exponential distributions are complete (Lehmann 1986). The toll (3.9b) may be negative, although it plus the delay D will be positive. The expected delay on finding n is n/ϕ_j, and this plus the toll (3.9a) gives the total $n(n+1)/2\phi_j$ proposed by Whittle (1985, 1986). It is interesting to note how rapidly these tolls grow with n or D; the quadratic functions (3.9) contrast markedly with the linear functions (3.8).

The multiclass network of $\cdot/M/1$ queues which has led to the forms (3.8) and (3.9) is a very special case, but it is certainly possible to extend the results further. We might, for example, want the delay through a queue to be distributed more like a normal than an exponential random variable. But a series of $m \cdot/M/1$ queues produces a delay D distributed as a gamma random variable with parameters m and $\phi_j-\rho_j$: for such a system

$$T_j(\rho_j) = m\rho_j(\phi_j-\rho_j)^{-2},$$

and

$$t_j(D) = \phi_j D^2/(m+1) - D, \quad D \geqslant 0,$$

is the unique function of d with expectation $T_j(\rho_j)$ for every traffic level ρ_j.

Throughout this section we have been concerned with adaptive routing; many interesting questions arise in connection with *dynamic* routing, where a user arriving at the network may have access to some or full information about the current network state (see, for example, Laws 1990, 1991). The distinction between adaptive and dynamic routing is not clearcut (see, for example, Kelly 1990), but one point is worth making here. If the system is striving to optimize the wrong function, then providing it with help, in the form of additional information, may damage performance. We illustrate this with an adaptation of Braess's paradox. Suppose the link between nodes 2 and 3 in figure 1*b* has a delay $\rho+10+X$, where X is a random variable taking the values 0 and 30 with equal probability. The random variable X may indicate the presence or absence of road works, for example. If users do not know the value of X then the expected delay along the link will be $\rho+25$, and the Wardrop equilibrium will be as in figure 1*a*. Each car incurs a delay of 83. If all users know the value of X then the equilibrium flow pattern will be either as in figure 1*a*, if $X=30$, or as in figure 1*b*, if $X=0$. The expected delay, averaged over the two possibilities, will be $\frac{1}{2}(83+92)=87\frac{1}{2}$. Providing all users with knowledge of X increases everyone's delay!

Current developments in electronics make feasible the practical application of both route guidance and road pricing (van Vuren & Smart 1990; Hoffman 1991) and the above discussion further emphasizes the importance of considering these topics together. If a system is encouraged to strive more aggressively towards an implicit objective, it becomes even more important that the objective is not perverse.

4. Loss networks

In this section we describe the basic theory of a loss network. The classical example is a telephone network, and we shall phrase our discussion in terms of calls, circuits and routes. However, the readers will observe that our model applies more widely to systems in which before a request (which may be a call, or a task, or a customer) is accepted it is first checked that sufficient resources are available to deal with the request.

Consider then a network with links from a set J, and suppose that link $j\in J$ comprises C_j circuits. A subset $r\subset J$ identifies a route. Calls requesting route r arrive as a Poisson stream of rate ν_r, and as r varies it indexes independent Poisson streams. A call requesting route r is blocked and lost if on any link $j\in r$ there is no free circuit. Otherwise the call is connected and simultaneously holds one circuit on each link $j\in r$ for the holding period of the call. The call holding period is randomly distributed with unit mean and independent of earlier arrival and holding times. Write R for the set of possible routes. Set $A_{jr}=1$ if $j\in r$, and $A_{jr}=0$ otherwise. This defines a $0-1$ matrix $A=(A_{jr}, j\in J, r\in R)$.

Let $n_r(t)$ be the number of calls in progress at time t on route r, and define the vectors $n(t)=(n_r(t), r\in R)$ and $C=(C_j, j\in J)$. Then the stochastic process $(n(t), t\geqslant 0)$ has a unique stationary distribution, and under this distribution $\pi(n)=P\{n(t)=n\}$ is given by

$$\pi(n)=G(C)^{-1}\prod_{r\in R}\frac{\nu_r^{n_r}}{n_r!},\quad n\in\mathscr{S}(C), \tag{4.1}$$

where

$$\mathscr{S}(C)=\{n\in\mathbb{Z}_+^R : An\leqslant C\}, \tag{4.2}$$

and $G(C)$ is the normalizing constant (or partition function)

$$G(C)=\left(\sum_{n\in\mathscr{S}(C)}\prod_{r\in R}\frac{\nu_r^{n_r}}{n_r!}\right). \tag{4.3}$$

This result is easy to check in the case where holding times are exponentially distributed: then $(n(t), t \geqslant 0)$ is a Markov process and the distribution (4.1) satisfies the detailed balance conditions

$$\nu_r \pi(n) = (n_r+1)\pi(n+e_r), \quad n, n+e_r \in \mathscr{S}(C),$$

where $e_r = (I[r' = r], r' \in R)$ is the unit vector describing just one call in progress on route r. In this form the result has been known for many years (see Brockmeyer *et al.* 1948): that the form (4.1) is insensitive to the holding time distributions is an example of the modern theory of insensitivity (see, for example, Whittle 1986).

Most quantities of interest can be written in terms of the distribution (4.1) or the partition function (4.3). For example let L_r be the proportion of calls requesting route r that are lost. Since the arrival stream of calls requesting route r is Poisson,

$$1-L_r = \sum_{n \in \mathscr{S}(C-Ae_r)} \pi(n) = G(C)^{-1} G(C-Ae_r). \tag{4.4}$$

Such simple explicit forms might be thought to provide the complete solution. However, this is far from the case. For all but the smallest networks it is impractical to compute G directly: observe that the number of routes $|R|$ may grow as fast as exponentially with the number of nodes $|J|$, and that in the (otherwise trivial) case when $|R| = |J|$ and $A = I$ the size of the state space $|\mathscr{S}(C)| = \Pi_{j \in J} C_j$ grows rapidly with the capacity limitations $C_j, j \in J$. The theory of computational complexity allows these remarks to be stated more formally. Louth (1991) has shown that, even in the restricted case where links have capacity 1 and arrival rates are equal, the task of computing the partition function (4.3) is #P-complete. Nevertheless, might a randomized algorithm, for example rejection sampling from the truncated Poisson distribution (4.1) or simulation of the underlying stochastic process, lead to an answer of any required accuracy, without the calculation growing exponentially with either the system size or the required accuracy? Again the answer is in general no: there can be no fully polynomial randomized approximation scheme for the task of computing the partition function (4.3), unless $RP = NP$ (Jerrum & Sinclair 1990; Louth 1991). Simulation is a valuable tool for many particular network structures, but these results indicate its limitations when dealing with arbitrary network topologies. A theme of much recent work has been to find approaches which complement computation and simulation with analytical insights.

Consider the problem of finding the most likely state n under the probability distribution (4.1). This is equivalent to maximizing

$$\sum_r (n_r \ln \nu_r - \ln n_r!),$$

over $n \in \mathscr{S}(C)$, a problem which is complicated by the discrete nature of the state space. To simplify things replace $\ln n!$ by $n \ln n - n$ (recall that by Stirling's formula $\ln n! = n \ln n - n + O(\ln n)$) and replace the integer vector n by a real vector x. The resulting problem is the following.

$$\text{Maximize} \quad \sum_r (x_r \ln \nu_r - x_r \ln x_r + x_r), \tag{4.5a}$$

$$\text{over} \quad x \geqslant 0, \tag{4.5b}$$

$$\text{subject to} \quad Ax \leqslant C. \tag{4.5c}$$

Observe that the objective function (4.5a) is differentiable and strictly concave over the cone $x \geqslant 0$ and tends to $-\infty$ as $\|x\| \rightarrow \infty$, and the feasible region (4.5c) is a closed convex set. Hence a maximizing value of x exists and is unique, and can be found by lagrangian methods (Whittle 1971). Consider, then, the lagrangian form

$$L(x, z; y) = \sum_r (x_r \ln \nu_r - x_r \ln x_r + x_r) + \sum_j y_j (C_j - \sum_r A_{jr} x_r - z_j)$$

$$= \sum_r x_r + \sum_r x_r (\ln \nu_r - \ln x_r - \sum_j y_j A_{jr}) + \sum_j y_j C_j - \sum_j y_j z_j,$$

where $z = (z_j, j \in J)$ is the vector of slack variables $z = C - Ax$ and $y = (y_j, j \in J)$ is a vector of Lagrange multipliers. To maximize $L(x, z; y)$ over the cone $x, z \geqslant 0$ we require that $y \geqslant 0, y \cdot z = 0$ and, differentiating with respect to x_r,

$$\ln \nu_r - \ln x_r - \sum_j y_j A_{jr} = 0.$$

The maximizing x_r is then

$$\bar{x}_r(y) = \nu_r \exp\left(-\sum_{j \in r} y_j\right), \tag{4.6}$$

and so

$$\max_{x, z \geqslant 0} L(x, z; y) = \sum_r \bar{x}_r(y) + \sum_j y_j C_j$$

$$= \sum_r \nu_r \exp(-\sum_{j \in r} y_j) + \sum_j y_j C_j.$$

Hence the lagrangian dual to the primal problem is the following.

$$\text{Minimize} \quad \sum_j \nu_r \exp(-\sum_{j \in r} y_j) + \sum_j y_j C_j, \tag{4.7a}$$

$$\text{over} \quad y \geqslant 0. \tag{4.7b}$$

We may solve the primal problem (4.5) by choosing values for the Lagrange multipliers $y = \bar{y}$ so that $\bar{x}(\bar{y}), \bar{y}$ are primal and dual feasible,

$$\bar{x}(\bar{y}) \geqslant 0, \quad \bar{z} = C - A\bar{x}(\bar{y}) \geqslant 0, \quad \bar{y} \geqslant 0 \tag{4.8a}$$

and satisfy the complementary slackness conditions

$$\bar{y} \cdot \bar{z}(\bar{y}) = 0. \tag{4.8b}$$

It is interesting to rewrite these conditions in terms of transformed variables

$$B_j = 1 - \exp(-y_j). \tag{4.9}$$

Under this transformation the conditions (4.8) on $\bar{y}$ become the following conditions on $B = (B_j, j \in J)$.

$$\sum_{r: j \in r} \nu_r \prod_{i \in r} (1 - B_i) = C_j \quad \text{if} \quad B_j > 0 \tag{4.10a}$$

$$\leqslant C_j \quad \text{if} \quad B_j = 0, \tag{4.10b}$$

$$B_1, B_2, \ldots, B_J \in [0, 1). \tag{4.10c}$$

The convexity properties of the primal problem (4.5) imply that there exist Lagrange multipliers $\bar{y}$ satisfying (4.8), and hence that there exists B satisfying (4.10). Alternatively, observe that the objective function of the dual problem (4.7a) is differentiable and convex over the cone $y \geqslant 0$ and tends to ∞ as $\|y\| \to \infty$. Hence an optimum $\bar{y}$ exists; differentiation of the dual objective function with respect to y_j

establishes a one-to-one correspondence under the transformation (4.9) between optima of the dual problem and solutions B to conditions (4.10). Finally observe that the mapping $y \mapsto yA$ is one-to-one from the set $y \geqslant 0$ if A has rank $|J|$. The objective function of the dual problem (4.7) is thus strictly convex if A has rank $|J|$. Hence the optimum $\bar{y}$ is unique if A has rank $|J|$.

In summary, we have shown that there exists a unique optimum to the primal problem (4.5), and that it can be expressed in the form

$$x_r = \nu_r \prod_{j \in r} (1 - B_j), \quad r \in R, \tag{4.11}$$

where $B = (B_j, j \in J)$ is any solution to the conditions (4.10) on B. There always exists a solution to the conditions on B, and it is unique if A has rank $|J|$. There is a one-to-one correspondence between solutions to the conditions (4.10) on B and optima of the dual problem (4.7), given by the transformation (4.9).

Conditions (4.10) have a straightforward interpretation in terms of a continuous, or fluid, flow. Suppose that an offered flow of ν_r on route r is thinned by a factor $(1-B_i)$ on each link $i \in r$ so that a flow of

$$\nu_r \prod_{i \in r} (1 - B_i) \tag{4.12}$$

remains. Then conditions (4.10) state that at any link j for which $B_j > 0$ the total capacity of that link, C_j, must be completely utilized by the superposition over routes r through link j of the flows (4.12). Conversely no thinning of flow is allowed at a link which is not full.

We have made some approximations in the formulation of the primal problem (4.5), and so the reader may well ask: What, precisely, is the connection between the simple form (4.11) and the distribution (4.1)? This connection we now outline.

The distribution (4.1) is that of $|R|$ independent Poisson random variables conditioned on a collection of linear inequality constraints (4.2). It is natural to look for a limit theorem as the capacities $C_j, j \in J$, and the offered traffics $\nu_r, r \in R$, are increased together (with ratios C_j/ν_r held fixed). We would expect the distribution (4.1) to approach that of $|R|$ independent normal random variables conditioned on a collection of linear inequality constraints. Limit results of this form are familiar when the linear inequalities are replaced by linear equalities, and arise naturally in the analysis of contingency tables (see, for example, Haberman 1974). The primal problem (4.5) and its solution (4.11) simply establish the form of the centring term in the expected central limit theorem (Kelly 1986; Hunt & Kelly 1989; Hunt 1990).

The central limit theorem implies, as a form of law of large numbers, that

$$L_r \rightarrow 1 - \prod_{j \in r} (1 - B_j), \quad r \in R, \tag{4.13}$$

where $B = (B_j, j \in J)$ is a solution to the dual problem (4.7). Here $1 - L_r$ is the exact acceptance probability, given by expression (4.4), and the limit is as capacities C and offered traffics ν are increased together, with ratios held fixed. Under this limiting régime it is *as if* links block independently, link j blocking with probability B_j.

The law of large numbers (4.13) has great appeal as a theoretical limit, but it has disappointing accuracy as a general approximation. In practice L_r is often approximated by the form

$$L_r \approx 1 - \prod_{j \in r} (1 - B_j), \quad r \in R, \tag{4.14}$$

but where $B_j, j \in J$, solve the nonlinear equations

$$B_j = E(\rho_j, C_j), \quad j \in J, \tag{4.15a}$$

$$\rho_j = \sum_{r:j\in r} \nu_r \prod_{i\in r-\{j\}} (1-B_i), \quad j \in J. \tag{4.15b}$$

Here the function $E(\cdot,\cdot)$ is defined for scalar ν and C by

$$E(\nu, C) = \frac{\nu^C}{C!}\left[\sum_{n=0}^{C} \frac{\nu^n}{n!}\right], \tag{4.16}$$

and is thus just Erlang's celebrated formula for the proportion of calls lost at single link of capacity C circuits offered Poisson traffic at rate ν. The idea underlying the approximation is simple to explain. Suppose that a Poisson stream of rate ν_r is thinned by factor $1-B_i$ at each link $i \in r-\{j\}$ before being offered to link j. If these thinnings could be assumed independent both from link to link and over all routes passing through link j (they clearly are not), then the traffic offered to link j would be Poisson at rate (4.15*b*), the blocking probability at link j would be (4.15*a*), and the loss probability on route r would satisfy (4.14) exactly.

A solution $B = (B_j, j \in J)$ to equations (4.15) exists, by the Brouwer fixed point theorem, but is it unique? We shall answer this question, and relate the approximation scheme (4.14)–(4.15) to our earlier limit results, by consideration of an appropriate optimization problem.

Define a function $U(y, C)$ by the implicit relation

$$U(-\ln(1-E(\nu, C)), C) = \nu(1-E(\nu, C)). \tag{4.17}$$

Observe that as ν increases from 0 to ∞ the first argument of U increases from 0 to ∞ and so this implicit relation defines a function $U: \mathbb{R}_+ \times \mathbb{Z}_+ \rightarrow \mathbb{R}_+$. Indeed, for a single link of capacity C circuits the quantity $U(y, C)$ is just the mean number of circuits in use (the *utilization*) when the blocking probability is $1-\exp(-y)$. Thus $U(y, C)$ is a strictly increasing function of y. Consider the following variant of the problem (4.7), which we shall call the revised dual problem.

$$\text{Minimize} \quad \sum_r \nu_r \exp\left(-\sum_{j\in r} y_j\right) + \sum_j \int_0^{y_j} U(z, C_j)\,\mathrm{d}z, \tag{4.18a}$$

$$\text{over} \quad y \geqslant 0. \tag{4.18b}$$

Since $U(y, C)$ is a strictly increasing function of y, $\int_0^y U(z, C)\,\mathrm{d}z$ is a strictly convex function of y. Hence the objective function (4.18*a*) is strictly convex, with a unique minimum. The objective function is also differentiable: hence the stationarity conditions

$$\sum_{r:j\in r} \nu_r \exp\left(-\sum_{i\in r} y_i\right) = U(y_j, C_j), \quad j \in J, \tag{4.19}$$

obtained by differentiating the objective function with respect to $y_j, j \in J$, locate a unique vector $y \geqslant 0$. Now suppose that $(B_j, j \in J)$ is a solution to equations (4.15). Under the equivalence $B = 1-\mathrm{e}^{-y}$ and using the definition (4.16), these equations become precisely equations (4.19). We deduce that equations (4.15) have a unique solution, given in terms of the optimum y of the problem (4.18) by the transformation (4.9).

The conditions (4.10) insist that the carried traffic on a link must equal capacity before the blocking probability on that link can be positive. Conditions (4.19) are a natural relaxation: as carried traffic approaches capacity, blocking increases, in a manner corresponding to the utilization function U. Note that replacing the function U by a function

$$U_f(z, C) = C, \quad z > 0,$$

reduces the revised dual problem (4.18) to the dual problem (4.7). The utilization function U_f would be natural for fluid flow, where if there is any blocking then all capacity is in use. For C large there is not much difference between U and U_f: under the limiting régime considered earlier, where C and ν increase together, the objective function (4.18*a*) approaches a scaled version of the objective function (4.7*a*). For non-limiting values of ν and C the approximation (4.14) is generally much improved when the vector B is defined, via the transformation (4.9), in terms of the optimum to the revised dual problem (4.18) rather than an optimum to the dual problem (4.7). This is intuitively plausible, since, as we have seen, the various equivalent formulations (4.15), (4.18) and (4.19) reflect the fact that as a link's utilization approaches its capacity, the link's blocking increases smoothly. As might be further expected from the informal motivation of the fixed point equations (4.15) in terms of thinned and superimposed Poisson streams, the approximation scheme (4.14)–(4.15) becomes increasingly accurate the more diverse the collection of routes passing through each link (for reviews, see Whitt 1985; Ziedinš 1987; Kelly 1991).

Thus the loss network, defined earlier in terms of Poisson arrival streams and certain rules for accepting calls, can also be viewed as a system implicitly attempting to solve the optimization problems (4.7) or (4.18). It is amusing to note that the term

$$\sum_r \nu_r \exp\left(-\sum_{j\in r} y_j\right) = \sum_r \nu_r \prod_{j\in r} (1-B_j),$$

appearing in both objective functions corresponds to the average level of traffic carried by the network; we would much prefer the network to be implicitly *maximizing* this term!

The loss network so far considered is a rather simple one, involving just *fixed* routing: if a call fails to be accepted on its fixed route then it is lost. In practice, networks often attempt to improve performance by allowing *alternative* routing, where a call which is blocked on a route may try again on an alternative route. We now consider a simple example of alternative routing in a fully connected network. Suppose that K nodes are linked to form a complete graph. Between any pair of nodes calls arise at rate ν, and there is a link of capacity C. If there is a spare circuit on the link joining the end points of a call then the call is accepted and carried by that circuit. Otherwise the call chooses at random a two-link path joining its end points: the call is accepted on that path if both links have a spare circuit, and is lost otherwise.

The generalization of equations (4.15) is based on the same underlying approximation, that links block independently. Let B be the link blocking probability, taken to be the same for each link. The probability that a call overflows from its first choice route is B, and the probability it can be accepted at the other link of a two-link alternative route is $1-B$; the arrival rate of overflowing calls at a link is then $2\nu B(1-B)$. We should look for a solution to the fixed point equation

$$B = E(\nu+2\nu B(1-B), C), \tag{4.20}$$

for B.

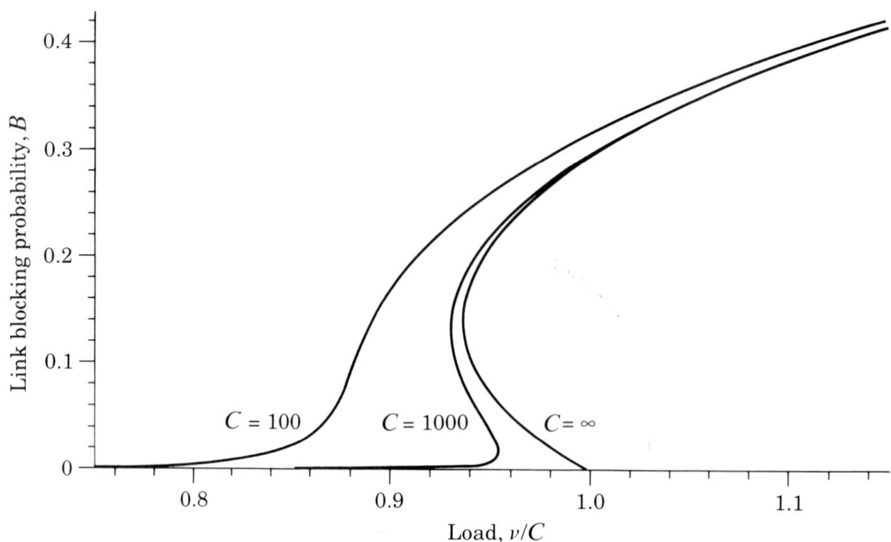

Figure 2. Instability of blocking probability.

The locus of points satisfying equation (4.20) is illustrated in figure 2. Observe the possibility of multiple solutions for B, when C is large enough and for a narrow range of the ratio ν/C. Are the multiple solutions evident in figure 2 a real phenomenon, or are they simply an artefact of the approximation? This question has been tackled by simulation of similar fully connected networks with tens or scores of nodes, and by rigorous analysis of the model as the number of nodes tends to infinity (Akinpelu 1984; Ackerley 1987; Gibbens *et al.* 1990; Crametz & Hunt 1991). The conclusions are clear. For networks with a moderate or large number of nodes the upper and lower solutions for B in figure 2 correspond to distinct, locally stable modes of a stationary distribution. There is also a hysteresis effect: if ν is varied slowly the mode which obtains may depend not just on the current value of ν but also upon whether ν approached this value from above or below. An intuitive explanation is easy to provide. The lower solution corresponds to a mode in which blocking is low, calls are mainly routed directly and relatively few calls are carried on two-link paths. The upper solution corresponds to a mode in which blocking is high and many calls are carried over two-link paths. Such calls use two circuits each, and this additional demand on network resources may cause a substantial number of subsequent calls also to attempt two-link paths. Thus a form of positive feedback may keep the system in the high blocking mode.

It is interesting to reinterpret this discussion in the language of catastrophe theory (Poston & Stewart 1978), as we now briefly indicate. By differentiating the potential function

$$\nu\,\mathrm{e}^{-y}+\nu\,\mathrm{e}^{-2y}(1-\tfrac{2}{3}\mathrm{e}^{-y})+\int_0^y U(z,C)\,\mathrm{d}z, \tag{4.21}$$

we find that it is stationary with respect to y when

$$\nu\,\mathrm{e}^{-y}+2\nu\,\mathrm{e}^{-2y}(1-\mathrm{e}^{-y}) = U(y,C). \tag{4.22}$$

However, under the equivalence $B = 1-\mathrm{e}^{-y}$ and using the definition (4.17), the equation (4.22) becomes precisely the equation (4.20). Thus the solutions illustrated in figure 2 locate the stationary points of the potential function (4.21). If we regard ν/C as the normal variable and C as the splitting variable, then figure 2 illustrates three cross sections of the cusp catastrophe.

Comparing the potential function (4.21) with the objective function (4.18*a*) of the revised dual problem, we see that alternative routing has led to the introduction of the second term in expression (4.21), and hence to non-convexity and multiple minima. The system is again implicitly minimizing a function, and bistability is to be expected when the function has two local minima.

For the parameter choice $(\nu, C) = (95, 100)$ the network loss probability, $B[1-(1-B)^2]$, takes a value of about 0.12. If alternative routing is not allowed, so that a call blocked on its direct link is lost, then the network loss probability is given by Erlang's formula (4.16) to be 0.05. Thus allowing a blocked call to attempt a two-link alternative route may *increase* the loss probability of the network. We have seen that this is plausible, since if a link accepts an alternatively routed call then it may later have to block a directly routed call, which will then attempt to find two circuits elsewhere in the network. A natural response is to allow a link to reject alternatively routed calls if the number of idle circuits on the link is less than or equal to a certain value, t, say. This method of giving priority at a link to certain traffic streams is known as *trunk reservation*, and the parameter t is called the trunk reservation parameter for the link. With trunk reservation in place alternative routing is capable of improving performance, and trunk reservation is widely used in telephone networks (Songhurst 1980).

5. Adaptive routing in loss networks

We have seen in the last section that the various potential functions implicitly minimized bear little direct relation to the network performance criteria of interest to system designers. In this section we show that an explicit consideration of such criteria can lead to decentralized adaptive routing schemes which are at least attempting to optimize the right function.

Suppose that each call carried on route r generates an expected revenue w_r (or, equivalently, interpret w_r as the cost of losing a call on route r). Then, under the fixed point model (4.15), the rate of return from the network will be

$$W(\nu; C) = \sum_r w_r \lambda_r,$$

where

$$\lambda_r = \nu_r \prod_{j \in r} (1-B_j),$$

corresponds to the traffic carried on route r. We use the notation $W(\nu; C)$ to emphasize the dependence of W on the vectors of offered traffics $(\nu_r, r \in R)$ and capacities $(C_j, j \in J)$. Let

$$\delta_j = \rho_j(E(\rho_j, C_j-1) - E(\rho_j, C_j)). \tag{5.1}$$

Extend the definition (4.16) to non-integral values of scalar C by linear interpolation, and at integer values of C_j define the derivative of $W(\nu; C)$ with respect to C_j to be the left derivative. Then it is possible to prove (Kelly 1988) that

$$\frac{\mathrm{d}}{\mathrm{d}\nu_r} W(\nu; C) = s_r \prod_{j \in r} (1-B_j), \tag{5.2}$$

and
$$\frac{\mathrm{d}}{\mathrm{d}C_j} W(\nu; C) = c_j, \tag{5.3}$$

where $s = (s_r, r \in R)$ and $c = (c_j, j \in J)$ are the unique solution to the linear equations

$$s_r = w_r - \sum_{j \in r} c_j, \tag{5.4}$$

$$c_j = \delta_j \sum_{r:j \in r} \lambda_r (s_r + c_j) \Big/ \sum_{r:j \in r} \lambda_r. \tag{5.5}$$

We can interpret s_r as the *surplus value* of a call on route r: if such a call is accepted it will earn w_r directly but at an *implied cost* of c_j for each circuit used from link j. The implied costs c measure the expected knock-on effects of accepting a call upon later arrivals at the network. From (5.3) it follows that c_j is also a *shadow price*, measuring the sensitivity of the rate of return to the capacity C_j of link j. The local character of equations (5.4) and (5.5) is striking. The right-hand side of (5.4) involves costs c_j only for links j on the route r, while (5.5) exhibits c_j in terms of an average, weighted over just those routes through link j, of $s_r + c_j$.

Expression (5.1) for δ_j is called Erlang's improvement formula (Brockmeyer *et al.* 1948). Observe that δ_j is simply the increase in the rate at which calls are blocked if a single link offered Poisson traffic at rate ρ_j has its capacity reduced by one circuit, and that δ_j increases from zero to one as ρ_j increases from zero to infinity.

Next we describe how equations (5.4) and (5.5) can be used as the basis for decentralized control of routing through the network. As motivation it is helpful to think in terms of the following model of a distributed computation. Suppose there is limited intelligence in the form of arithmetical processing ability available for each link j and for each route r. This intelligence may be located centrally or it may be distributed over the nodes of the network; for example the processing for route r might be carried out by the source node for calls on route r. Suppose also that there is the possibility of limited communication between the intelligences of link j and route r provided $j \in r$. Consider now equations (5.4) and (5.5). One method for attempting a solution to these equations is repeated substitution. Choose a vector c; substitute it in equation (5.4) to obtain a vector s; substitute these into equation (5.5) to obtain a revised vector c, and repeat. This computation can, however, be distributed over the intelligences of links and routes, since equation (5.4) for s_r involves implied costs c_j only for links j on the route r, while equation (5.5) for c_j involves only surplus values s_r for routes r passing through link j. Kelly (1988) showed that if

$$\sum_{i \in r - \{j\}} \delta_i < 1, \quad j \in r, \quad r \in R, \tag{5.6}$$

then repeated substitution converges to the unique solution of the equations. When the condition (5.6) is not satisfied it may be necessary to damp the repeated substitution to obtain convergence. This can still be implemented by a distributed computation, but the individual intelligences may require some knowledge of the network beyond that locally available, to damp sufficiently the repeated substitution. In fact it is enough for the intelligences to know just one item of global information, namely J, the total number of links in the network.

The quantities δ_j and λ_r appearing in equations (5.4) and (5.5) are not fixed and known. However, they can be estimated by intelligences of links and routes from, for example, local measurements of carried loads. The estimates can then be used in a distributed computation of the vector s. Finally the derivatives (5.2) can be used to implement a decentralized hill-climbing search procedure able to adapt routing patterns in response to changes in the demands on the network.

The inequality (5.6) limits the size of knock-on effects and is a form of light traffic condition. If it is satisfied then the fixed point model (4.15) has a certain stability property: any perturbation of the capacity C_j of link j may cause a change in the entire vector $(\lambda_r, r \in R)$ of carried traffics, but the change in component λ_r diminishes rapidly with the extent of the separation between route r and link j. Without condition (5.6) perturbations may have influence over arbitrarily great distances. The existence of such effects is *not* an artefact of the approximation: examples have been deduced from the exact stationary distribution (4.1). The discussion of these effects is complicated by *frustration*, the crucial property that lies at the heart of the spin glass problem of statistical mechanics, whereby chains of influence throughout the network may be out of phase and compete with one another. The reader is referred to Kelly (1991) for further discussion and references.

It is possible to define implied costs and surplus values for fixed point models of alternative routing and trunk reservation, and to show that they solve linear relations generalizing (5.4) and (5.5) (see Key 1988; Key & Whitehead 1988; Kelly 1990). The potential for long-range order and instability is more pronounced in networks with alternative routing, since the chains of influence along different paths tend to reinforce one another. The possibility emphasizes the importance of treating a network as a whole: the local benefits of a capacity or routing change may be completely overwhelmed by adverse consequences elsewhere in the network.

6. Dynamic Alternative Routing

The routing scheme discussed in the last section attempts to control the network by explicitly computing average traffics, implied costs and surplus values, and deducing derivatives of performance criteria. The scheme resembles the adaptive schemes of §3, and indeed the implied costs precisely parallel the tolls discussed here. Might it be possible to design a network so that it instead resembles the electrical network of §2, with simple rules for call acceptance and routing leading naturally to good behaviour? Early work on this question has led to a scheme, Dynamic Alternative Routing (DAR), now being implemented in British Telecom's British trunk network.

DAR is a simple but effective dynamic routing strategy, which is decentralized and uses only local information. Its definition for a fully connected network is as follows. Suppose there are K nodes in the network, with the link $\{i,j\}$ joining nodes i and j having capacity C_{ij}. Each link is assigned a trunk reservation parameter t_{ij}, and each source–destination pair (i,j) stores the identity of its current tandem $k(i,j)$ for use in two-link alternative routes. A call between nodes i and j is first offered to the direct link and a call is always routed along that link if there is a free circuit. Otherwise, the call attempts the two-link alternative route via tandem node k with trunk reservation applied to both links. If the call fails to be routed via k, this call is lost and, further, the identity of the tandem node is reselected (at random perhaps) from the set $\{1, 2, \ldots, K\} - \{i,j\}$. Note especially that the tandem node is not reselected if the call is successfully routed on either the direct link or the two-link alternative route. Mees (1986) has coined the term *sticky random routing* to emphasize this property of the scheme. In practice it has been found simpler to reselect a tandem node by cycling around a fixed random permutation; the point is that reselection is not based on any collected data, only the important information that a call has just failed.

Let $p_k(i,j)$ denote the long-run proportion of calls between i and j which are offered to tandem node k, and let $q_k(i,j)$ be the long-run proportion of those calls between i and j and offered to tandem node k which are blocked. Then, under uniform reselection,

$$p_a(i,j)\,q_a(i,j) = p_b(i,j)\,q_b(i,j), \quad a,b \neq i,j.$$

Observe that this simple ergodic result is exact for either random reselection or reselection using a fixed permutation. More generally, suppose the DAR mechanism for reselection of the tandem node between i and j chooses node k with long-run frequency f_k, where $\Sigma_{k\neq i,j} f_k = 1$. Then each selection of node k is paired with a failed call via node k, and so

$$p_a(i,j)\,q_a(i,j) : p_b(i,j)\,q_b(i,j) = f_a : f_b, \quad a,b \neq i,j. \tag{6.1}$$

Observe that if the blocking $q_k(i,j)$ is high on the path through the tandem node k, then the proportion of overflow routed via node k will be low. This gives some insight into the means by which the routing scheme implicitly adapts to overloads and failures.

For further insight consider next the effects of mismatches between traffics and capacities. Suppose that traffic is greater than capacity on some links, and less than capacity on others. Is it possible for the excess traffic from overloaded links to be assigned to alternative routes which do not themselves clash with one another; and, if so, is it possible for this to be achieved by a simple algorithm? To proceed with these questions, consider the following random graph problem. Choose $p \in (0,1)$ and suppose the edges of a fully connected graph on K nodes are independently coloured red with probability p and white with probability $1-p$. Let a *triangle* be a set of three edges joining each pair from a set of three nodes. Call a triangle *good* if it contains one red and two white edges. Let $P(K)$ be the probability that there exists a set of disjoint good triangles such that each red edge is contained in a triangle. If we interpret the red edges as overloaded links, then $P(K)$ is the probability that there exists a collection of non-overlapping alternative routes. Hajek (1987) has shown that if $p < \frac{1}{3}$ then $P(K) \to 1$ as $K \to \infty$. Hajek's methods are informative about the performance of algorithms as well as the structure of random graphs, and we outline one aspect of his proof. Consider the following very simple greedy algorithm. Suppose disjoint good triangles $T_1, T_2, \ldots, T_k$ have been found already and the algorithm has not yet stopped. If there is no remaining red edge, declare the algorithm successful and stop. Otherwise, call a triangle available (after k steps) if it is a good triangle which is disjoint from $T_1, T_2, \ldots, T_k$. Choose a red edge e at random from the remaining red edges. If no available triangle contains e, declare the algorithm unsuccessful and stop. Otherwise, choose T_{k+1} at random from among the available triangles containing e. Hajek conjectures that this simple greedy algorithm is successful with probability approaching 1. He proves that when the algorithm stops it has covered at least a proportion $1-\delta$ of the red edges with probability approaching 1 as $K \to \infty$, for any $\delta > 0$. He establishes the above result by consideration of a modified algorithm. Choose ϵ with $0 < \epsilon < \frac{1}{3} - p$. Independently consider each white edge and delete it with probability ϵ. Now run the original algorithm (although now only available triangles with non-deleted edges are counted). Then this modified algorithm is successful with probability approaching 1.

It is informative that a simple greedy algorithm performs so well for the above random graph problem. In many respects DAR resembles a dynamic version of the greedy algorithm. It also has a number of similarities with probabilistic hill-climbing techniques such as simulated annealing, and Gibbens (1988) and Mitra & Seery (1991) have discussed its use as an algorithm to investigate random graph and network routing problems. Consider the multi-commodity flow problem of routing $n(n-1)$ distinct flows between each ordered pair of n nodes. As a linear program this problem has $O(n^3)$ variables, since each of $n(n-1)$ ordered node pairs has one direct route and $(n-2)$ two-link alternatives available, and $O(n^2)$ constraints, associated with the $\frac{1}{2}n(n-1)$ links and distinct flows (cf. Gibbens & Kelly 1990). The problem has a natural mapping onto a computing engine with $O(n^2)$ parallel processors. DAR tackles the nonlinear stochastic version of this problem using a probabilistic hill-climbing technique performed by $O(n^2)$ parallel processors, namely the occupancy levels of the $\frac{1}{2}n(n-1)$ links and the $n(n-1)$ tandem pointers $k(i,j)$.

The formula (6.1) allows the fixed point methods described in §4 to be extended to model DAR; these methods, together with simulation, have been used to investigate the performance of DAR under a wide range of failure and overload conditions. For an account of this work, and extensions of DAR to regular but non-fully connected networks, the reader is referred to Stacey & Songhurst (1987), Gibbens (1988), Gibbens *et al.* (1988), Gibbens & Kelly (1990), Gibbens & Turner (1991).

The routing scheme described in §5 has the advantage that it makes no special assumption about the underlying topology of the network, but it adapts slowly, in response to averaged values. DAR, in comparison, requires some regularity of network structure, but operates on a fast timescale, driven by call arrivals, and uses very simple rules. Because of the different timescales involved it is quite possible that the benefits of both schemes may be achievable: certainly Key (1988) has described an approach to the calculation of implied costs and shadow prices for a fully connected network using DAR, and has shown how the shadow prices can be used interactively to allocate capacity in a network. Current research is developing methods which use sticky random routing on a fast timescale, driven by call arrivals or failure events, with route sets, priorities and trunk reservation selected using implied costs calculated over the longer timescales necessary for averages to be estimated.

This paper is an expanded version of a talk first given at the Annual Meeting of the British Association for the Advancement of Science, in Sheffield, 1989. The author gratefully acknowledges the support of the Science and Engineering Research Council, under grant GR/E83009.

References

Ackerley, R. G. 1987 Hysteresis-type behaviour in networks with extensive overflow. *Br. Telecom Technol. J.* **5**, 42–50.

Akinpelu, J. M. 1984 The overload performance of engineered networks with non-hierarchical routing. *AT&T Tech. J.* **63**, 1261–1281.

Anderson, P. W., Arrow, K. J. & Pines, D. (eds) 1988 *The economy as an evolving complex system.* Redwood City, California: Addison-Wesley.

Braess, D. 1968 Über ein Paradoxon aus der Verkehrsplanung. *Unternehmenforschung* **12**, 258–268.

Brockmeyer, E., Halstrom, H. L. & Jensen, A. 1948 *The life and works of A. K. Erlang.* Copenhagen: Academy of Technical Sciences.

Cohen, J. E. 1988 The counterintuitive in conflict and cooperation. *Am. Scient.* **76**, 576–584.

Cohen, J. E. & Kelly, F. P. 1990 A paradox of congestion in a queueing network. *J. appl. Prob.* **27**, 730–734.

Crametz, J.-P. & Hunt, P. J. 1991 A limit result respecting graph structure for a fully connected network with alternative routing. *Ann. Appl. Prob.* **1**.

Doyle, P. G. & Snell, J. L. 1984 *Random walks and electric networks.* Mathematical Association of America.

Erlang, A. K. 1925 A proof of Maxwell's law, the principal proposition in the kinetic theory of gases. In Brockmeyer *et al.* (1948), pp. 222–226.

Gallager R. G. 1977 A minimum delay routing algorithm using distributed computation. *IEEE Trans. Commun.* **25**, 73–85.

Gibbens, R. J. 1988 Dynamic routing in circuit-switched networks: the dynamic alternative routing strategy. Ph.D. thesis, University of Cambridge.

Gibbens, R. J., Kelly, F. P. & Key, P. B. 1988 Dynamic alternative routing – modelling and behaviour. *Proc. 12th Int. Teletraffic Congress, Turin* (ed. M. Bonatti). Amsterdam: Elsevier.

Gibbens, R. J., Hunt, P. J. & Kelly, F. P. 1990 Bistability in communication networks. In *Disorder in physical systems* (ed. G. R. Grimmett & D. J. A. Welsh), pp. 113–128. Oxford: Oxford University Press.

Gibbens, R. J. & Kelly, F. P. 1990 Dynamic routing in fully connected networks. *IMA J. Math. Cont. Information* **7**, 77–111.

Gibbens, R. J. & Turner, S. R. E. 1991 Sticky random routing in dual-parented networks. *Eighth UK Teletraffic Symposium, Nottingham.* London: IEE.

Haberman, S. J. 1974 *The analysis of frequency data.* University of Chicago Press.

Hajek, B. 1987 Average case analysis of greedy algorithms for Kelly's triangle problem and the independent set problem. In *26th IEEE Conference on Decision and Control.* New York: IEEE.

Hoffman, G. 1991 Up-to-the-minute information as we drive – how it can help road users and traffic management. *Transport Rev.* **11**, 41–61.

Hunt, P. J. 1990 Limit theorems for stochastic loss networks. Ph.D. thesis, University of Cambridge.

Hunt, P. J. & Kelly, F. P. 1989 On critically loaded loss networks. *Adv. appl. Prob.* **21**, 831–841.

Jerrum, M. R. & Sinclair, A. 1990 Polynomial time approximation algorithms for the Ising model. Department of Computer Science, Edinburgh.

Kelly, F. P. 1979 *Reversibility and stochastic networks.* Chichester: Wiley.

Kelly, F. P. 1986 Blocking probabilities in large circuit-switched networks. *Adv. appl. Prob.* **18**, 473–505.

Kelly, F. P. 1988 Routing in circuit-switched networks: optimization, shadow prices and decentralization. *Adv. appl. Prob.* **20**, 112–144.

Kelly, F. P. 1990 Routing and capacity allocation in networks with trunk reservation. *Math. oper. Res.* **15**, 771–793.

Kelly, F. P. 1991 Loss networks. *Ann. appl. Prob.* **1**, 319–378.

Key, P. B. 1988 Implied cost methodology and software tools for a fully connected network with DAR and trunk reservation. *Br. Telecom Technol. J.* **6**, 52–65.

Key, P. B. & Whitehead, M. J. 1988 Cost-effective use of networks employing Dynamic Alternative Routing. In *Proc. 12th Int. Teletraffic Congress, Turin* (ed. M. Bonatti). Amsterdam: Elsevier.

Kingman, J. F. C. 1969 Markov population processes. *J. appl. Prob.* **6**, 1–18.

Knödel, W. 1969 *Graphentheoretische Methoden und ihre Anwendungen*, pp. 56–59. Berlin: Springer-Verlag.

Langton, C. G. (ed.) 1989 *Artificial life.* Redwood City, California: Addison-Wesley.

Laws, C. N. 1990 Dynamic routing in queueing networks. Ph.D. thesis, University of Cambridge.

Laws, C. N. 1992 Resource pooling in queueing networks with dynamic routing. *Adv. appl. Prob.* **24**. (In the press.)

Lehmann, E. L. 1986 *Testing statistical hypotheses*, 2nd edn. New York: Wiley.

Louth, G. M. 1991 Stochastic networks: complexity, dependence and routing. Ph.D. thesis, University of Cambridge.

Mees, A. I. 1986 Simple is the best for dynamic routing of telecommunications. *Nature, Lond.* **323**, 108.

Mitra, D. & Seery, J. B. 1991 Comparative evaluations of randomized and dynamic routing strategies for circuit-switched networks. *IEEE Trans. Commun.* **39**, 102–116.

Nagurney, A. 1987 Competitive equilibrium problems, variational inequalities and regional science. *J. Regional Sci.* **27**, 503–517.

New York Times 1990 What if they closed 42nd Street and nobody noticed? 25 December, p. 38.

Pigou, A. C. 1920 *The economics of welfare.* London: Macmillan.

Pines, D. (ed.) 1987 *Emerging synthesis in science.* Redwood City, California: Addison-Wesley.

Poston, T. & Stewart, I. 1978 *Catastrophe theory and its applications.* London: Pitman.

Potts, R. B. & Oliver, R. M. 1972 *Flows in transportation networks.* New York: Academic.

Songhurst, D. J. 1980 Protection against traffic overload in hierarchical networks employing alternative routing. *Telecommunications Network Planning Symposium, Paris.*

Stacey, R. R. & Songhurst, D. J. 1987 Dynamic Alternative Routing in the British Telecom trunk network. *Int. Switching Symposium, Phoenix, Arizona.*

Thomson, W. & Tait, P. G. 1879 *Treatise on natural philosophy.* Cambridge.

van Vuren, T. & Smart, M. B. 1990 Route guidance and road pricing – problems, practicalities and possibilities. *Transport Rev.* **10**, 269–283.

Walters, A. A. 1961 The theory and measurement of private and social cost of highway congestion. *Econometrica* **29**, 676–699.

Wardrop, J. G. 1952 some theoretical aspects of road traffic research. *Proc. Inst. Civil Engng* **1**, 325–378.

Whitt, W. 1985 Blocking when service is required from several facilities simultaneously. *AT&T Tech. J.* **64**, 1807–1856.

Whittle, P. 1971 *Optimization under constraints.* London: Wiley.

Whittle, P. 1985 Scheduling and characterization problems for stochastic networks (with Discussion). *Jl R. statist. Soc.* **47**, 407–428.

Whittle, P. 1986 *Systems in stochastic equilibrium.* Chichester: Wiley.

Wroe, G. A., Cope, G. A. & Whitehead, M. J. 1990 Flexible routing in the BT international access network. *Seventh UK Teletraffic Symposium, Durham.* London: IEE.

Ziedinš, I. B. 1987 Stochastic models of traffic in star and line networks. Ph.D. thesis, University of Cambridge.

Bayesian computational methods

By Adrian F. M. Smith

Department of Mathematics, Imperial College of Science, Technology and Medicine, London SW7 2BZ, U.K.

The bayesian (or integrated likelihood) approach to statistical modelling and analysis proceeds by representing all uncertainties in the form of probability distributions. Learning from new data is accomplished by application of Bayes's Theorem, the latter providing a joint probability description of uncertainty for all model unknowns. To pass from this joint probability distribution to a collection of marginal summary inferences for specified interesting individual (or subsets of) unknowns, requires appropriate integration of the joint distribution. In all but simple stylized problems, these (typically high-dimensional) integrations will have to be performed numerically. This need for efficient simultaneous calculation of potentially many numerical integrals poses novel computational problems. Developments over the past decade are reviewed, including adaptive quadrature, adaptive Monte Carlo, and a variant of a Markov chain simulation procedure known as the Gibbs sampler.

1. Introduction

The bayesian (or integrated likelihood) approach to statistical modelling proceeds by specifying a probability model, $p(x \mid \theta)$, for data realizations, x, together with an *a priori* (prior) probability weighting distribution, $p(\theta)$, for possible values of the unknown model parameters, θ. Statistical analysis (i.e. making inferences about unknown parameters, or predicting future data values) is then based on $p(\theta \mid x)$, the *a posteriori* (posterior) uncertainty distribution for unknown parameters (conditional on observed data), given by Bayes's Theorem,

$$p(\theta \mid x) = p(x \mid \theta)\, p(\theta) \Big/ \int p(x \mid \theta)\, p(\theta)\, \mathrm{d}\theta, \tag{1.1}$$

and the (predictive) uncertainty distribution for future data y generated by $p(y \mid \theta)$ is given by

$$p(y \mid x) = \int p(y \mid \theta)\, p(\theta \mid x)\, \mathrm{d}\theta.$$

In formal terms, this bayesian inference prescription is deceptively simple: we just specify $p(x \mid \theta)$ (usually referred to as the likelihood, $l(\theta; x)$, when considered as a function of θ) and $p(\theta)$, multiply the likelihood and prior probability functions together and normalize to form the posterior probability function.

From this joint posterior density over all unknown model parameters, by standard probability marginalization and transformation techniques we can straightforwardly obtain any particular univariate or bivariate, etc., summary in the form of densities, contours or moments as required.

However, calculation of the joint density for θ, together with the required marginalization and moment summaries, rests on an ability to perform a number of

Phil. Trans. R. Soc. Lond. A (1991) **337**, 369–386

Printed in Great Britain

(typically high-dimensional) integrations. In particular, it is necessary to find the normalizing constant of the full joint density and to eliminate complementary components of θ, or transformations of θ, to obtain marginal densities or marginal moment summaries. The technical problem of carrying out a bayesian analysis therefore reduces to that of performing or approximating a number of (typically high-dimensional) numerical integrals.

In practice, the mathematical forms of $l(\theta;x)$ and/or $p(\theta)$ make this process far from trivial, as we shall try to indicate by the following brief comments on a range of models and applications.

(*a*) *Image analysis*

Problems of image analysis, reconstruction, edge and feature detection, etc., arise throughout the sciences, medicine and technology. In these contexts, θ is the unknown true image, the θ-vector components being the true individual pixel values. The data vector x is the observed (noisy) image, with $p(x\,|\,\theta)$ describing the noise process and taking characteristically different forms (often the result of an intricate modelling process) in different application areas. The prior specification $p(\theta)$ characterizes (aspects of) the underlying true image and could take the form of a detailed probabilistic template, or a stylized statement about local correlation. Clearly, challenging computational problems arise, the dimension of θ often being of the order of 256×256 (and perhaps even 1000×1000).

(*b*) *Mixture analysis*

Titterington *et al.* (1986) detail numerous application areas where the data model, $p(x\,|\,\theta)$, is a weighed linear combination (a finite mixture) of component models. Thus, for example, x may be observed fish length, from a given species, and the component models correspond to length distributions at different fish ages. The unknown mixture weights are the proportions of different age groups in the underlying fish population; the unknown parameters of the component models characterize mean size and variability at different ages. Here, the prior specification, $p(\theta)$, needs to describe the relationships among these latter parameters, as well as information on age distribution. In the case of fisheries research, inference may focus on the mixture proportions; in other applications of mixtures, interest may focus on the component model parameters, or on deciding the number of components. Computational problems in mixture analysis arise both from potentially high dimensionality of θ (for example, a 12-component bivariate gaussian mixture has 71 unknown parameters) and the notoriously complex form of the likelihood function.

(*c*) *Change-point analysis*

Change-point models arise when a mechanism or structure (e.g. biological, economic or social) is characterized by periods of stability (although subject to statistical variability), but with sudden shifts from one stable form to another. Examples are provided by growth phases in biology, shifts in economic behaviour, archaeological phases corresponding to abandonment and resettlement of sites, and changes in clinical conditions, such as alternating periods of rejection and recovery in organ transplantation. The data model, $p(x\,|\,\theta)$, has components for each stable phase, with θ representing parameters for each component, as well as the unknown change-points (which, typically, refer to time). The prior specification needs to represent information about the change mechanism, as well as relationships among parameters within and between phases. Again, dimensionality quickly becomes a problem (for example, a heteroscedastic segmented multiple regression with 10 regressors and up to five phases has at least 64 unknown parameters) and there are often awkward parametric constraints (for example, if segmented response curves are assumed continuous at change-points). In many applications, inference focuses on the change-points themselves. In some cases, for example economic forecasting, interest may focus on inference or prediction based just on the identified final data régime.

(*d*) *Hierarchical analysis*

Hierarchical models, often referred to under the label of Empirical Bayes models, arise as follows. The data model, $p(x \mid \theta)$, is a combination of data models, $p_i(x_i \mid \theta_i)$, from a number of separate contexts, and $p(\theta)$ has the structure

$$p(\theta) = \int p(\theta \mid \phi)\, p(\phi)\, \mathrm{d}\phi,$$

where $p(\theta \mid \phi)$ is a further layer of modelling describing an assumed relationship among the θ_i (with the 'hyperparameter', ϕ, functioning as an unknown parameter in $p(\theta \mid \phi)$). Examples include: population modelling in the pharmaceutical sciences, where the $p_i(x_i \mid \theta_i)$ model drug concentration profiles for individuals i, having kinetic and measurement noise parameters θ_i, with $p(\theta)$ modelling the variability of individual kinetic parameters in the population, as well as information about measurement noise; spatial modelling in epidemiology, where the $p_i(x_i \mid \theta_i)$ model disease incidence rates in different locations, with $p(\theta)$ modelling the spatial and other variability of underlying factors. Dimensionality again becomes a problem: for example, if drug concentration profiles for 50 individuals are modelled by a four-parameter function with additive gaussian noise, the combined vector of unknowns (θ, ϕ) may be 264-dimensional (even more if models are extended to deal with potential outliers or the need for data transformation). In many applications, interest focuses on the population characteristics, described by ϕ. In other cases, the individual θ_i (or predictions for individuals or locations) are of equal interest.

This brief summary of a selection of typical problem areas in which bayesian modelling and analysis is used serves, we hope, to indicate and illustrate the very real computational challenges posed by the dimensionality and complexity arising from realistic forms of $p(x \mid \theta)$ and $p(\theta)$. The remainder of this paper presents a review of some of the approaches developed over the past decade in response to these computational challenges.

In §2, we review various analytic approximation strategies that have been proposed. In §3, we outline the ways in which a bayesian response to the problem of simultaneously performing a number of high-dimensional numerical integrals suggests novel iterative, adaptive strategies based on mixes of cartesian product, spherical and importance sampling quadrature techniques. In §4, we outline recent ideas based on sampling and resampling strategies. In §5, we provide a brief comparative overview of the various techniques reviewed.

2. Analytic approximation

For completeness, we begin with a brief review of familiar analytic approaches to approximating posterior inference summaries.

(a) The assumption of asymptotic normality

If $\ln l(x;\theta)$ is denoted by $L(\theta)$, suppressing the dependence on the given data x, it is well known that, for large sample sizes, the posterior distribution of θ is often approximately $N(\hat{\theta}, \hat{\Sigma})$, where $\hat{\theta}$ is the maximum likelihood estimate and $\hat{\Sigma}$ is the inverse of the hessian matrix evaluated at $\hat{\theta}$.

This approach has the powerful advantage that practically all forms of summary posterior inference are trivially calculated as by-products of the normality assumption. Moreover, computer programs are readily available for calculating $\hat{\theta}$ and $\hat{\Sigma}$: indeed, apart from differences in philosophical approach and interpretation, these latter quantities are precisely those widely used as the basis for summary inferences by proponents of likelihood inference.

However, there is a major problem with this strategy and one that receives far too little attention in practice, namely, how to check, in any specific application, that the assumption of approximate normality is really justified.

(b) Lindley's approximations

Lindley (1980) develops specific expansions to order n^{-1} for ratios of integrals of the form

$$\int w(\theta)\, l(x;\theta)\, \mathrm{d}\theta \Big/ \int v(\theta)\, l(x;\theta)\, \mathrm{d}\theta. \tag{2.1}$$

Thus, for example, with $w(\theta) = g(\theta)\, p(\theta), v(\theta) = p(\theta), \rho(\theta) = \ln p(\theta)$, Lindley shows that

$$E(g(\theta)\,|\,x) \approx g + \tfrac{1}{2}\sum_{i,j} (g_{ij} + 2g_i \rho_i)\, \sigma_{ij} + \tfrac{1}{2} \sum_{i,j,l,m} L_{ijm}\, g_l\, \sigma_{ij}\, \sigma_{ml}, \tag{2.2}$$

where subscripts on g, ρ and L denote differentiation with respect to specific components of θ, with all functions evaluated at $\theta = \hat{\theta}$, and the σ_{ij} are the elements of $\hat{\Sigma}$.

Lindley's expansions provide, in a sense, first-order corrections to the maximum likelihood approximations which incorporate some influence from the prior distribution. Unfortunately, however, the approximations require the evaluations of up to third-order derivatives, a task which quickly becomes irksome as the parameter dimension k increases.

(c) The Laplace approximation

More recently, Tierney & Kadane (1986) proposed a form of analytic approximation, which requires the evaluation of only first and second derivatives of slightly modified likelihood functions. The basic idea is to apply separately the Laplace method for integrals to both the numerator and denominator in (2.1). We can illustrate the method by considering again the special case $w(\theta) = g(\theta)\, p(\theta)$, $v(\theta) = p(\theta)$, for positive $g(\cdot)$. We take

$$M(\theta) = n^{-1}(L(\theta) + \ln p(\theta)) \approx M(\hat{\theta}_M) - (\theta - \hat{\theta}_M)^2/(2\sigma_M^2), \tag{2.3}$$

where $\hat{\theta}_M$ maximizes $M(\theta)$ and σ_M^2 is minus the inverse of the second derivative of $M(\cdot)$ evaluated at $\hat{\theta}_M$, and then approximate the denominator of (2.1) by

$$\int l(x;\theta)\, p(\theta)\, \mathrm{d}\theta = \int \mathrm{e}^{nM(\theta)}\, \mathrm{d}\theta \approx \surd(2\pi)\, \sigma_M\, n^{-\frac{1}{2}} \exp\{nM(\hat{\theta}_M)\}. \tag{2.4}$$

To approximate the numerator, we take

$$N(\theta) = n^{-1}\{\ln g(\theta) + L(\theta) + \ln p(\theta)\}$$

and evaluate the maximum $\hat{\theta}_N$ of $N(\theta)$, together with σ_N^2, which is equal to minus twice the second derivative of $N(\cdot)$ evaluated at $\hat{\theta}_N$, obtaining

$$\int g(\theta)\, l(x;\theta)\, p(\theta)\, \mathrm{d}\theta \approx \sqrt{(2\pi)}\, \sigma_N\, n^{-\frac{1}{2}} \exp\{nN(\hat{\theta}_N)\}. \tag{2.5}$$

From (2.4) and (2.5) we obtain the approximation

$$E[g(\theta)\,|\,x] \approx (\sigma_N/\sigma_M) \exp\{n[N(\hat{\theta}_N) - M(\hat{\theta}_M)]\}, \tag{2.6}$$

which has relative error of order n^{-2} and seems to be more accurate than conventional approximations for a range of problems.

The above developments have obvious direct application to the calculation of posterior moments and predictive densities. In addition, Tierney & Kadane extend these ideas to multiparameter situations and employ Laplace's method to perform the approximate integration of subsets of components of the parameter vector from the overall joint posterior density in order to obtain marginal joint posterior densities. Recent extensions and refinements of the methodology include Kass *et al.* (1989) and Tierney *et al.* (1989). An alternative analytic approximation technique is discussed by Leonard *et al.* (1989).

3. Numerical integration

(*a*) *Background*

As noted in Naylor & Smith (1982), the numerical calculation and marginalization of $p(\theta\,|\,x)$, as given in (1.1) above, involves the computation of integrals of the form

$$S_I[q(\theta)] = \int q(\theta)\, l(x;\theta)\, p(\theta)\, \mathrm{d}\theta_{I'}, \tag{3.1}$$

where, if $\theta_1, \ldots, \theta_k$ denote the components of θ, I is some index set $I \subseteq \{1, \ldots, k\}$ and $\theta_{I'}$, denotes the vector of components of θ whose subscripts are not elements of I. Thus for example, $I = \varnothing$, $q(\theta) = 1$ corresponds to the normalizing constant in (1.1), $S_I[1]/S_\varnothing[1]$ corresponds to the marginal posterior density ordinate $p(\theta_i\,|\,x)$, and $S_\varnothing[\theta_i \theta_j]/S_\varnothing[1]$ corresponds to the posterior expectation of $\theta_i \theta_j$. Other moments and predictive densities, etc., are similarly obtained from integrals having the form (3.1).

In the case of the k-dimensional integrals,

$$S[q(\theta)] = \int q(\theta)\, l(x;\theta)\, p(\theta)\, \mathrm{d}\theta, \tag{3.2}$$

it is well known that efficient quadrature formulae are available if the integrand in (3.2) can be well approximated by a function of the form

$$g(\theta) = h(\theta)\, n(\theta;\phi,\Sigma), \tag{3.3}$$

where $n(\theta;\phi,\Sigma)$ denotes a k-dimensional normal density, with known mean ϕ and known covariance matrix Σ, and $h(\theta)$ is a polynomial in the components of θ. In such cases, a constructive orthogonalizing and centring and scaling transformation of the form

$$\psi_1 = \theta_1,$$

$$\psi_i = \theta_i + \sum_{j=1}^{i-1} \beta_{ij} \psi_j, \quad i = 2, \ldots, k, \tag{3.4}$$

with
$$\beta_{ij} = -\mathrm{cov}\,(\theta_i, \psi_j\,|\,x)/\mathrm{var}\,(\psi_j\,|\,x),$$
followed by
$$\xi_i = (\psi_i - \mu_i)/(\surd 2\sigma_i), \tag{3.5}$$
where μ_i, σ_i^2 are the means and variances of the ψ_i, leads to the standardized form
$$\int g(\theta)\,\mathrm{d}\theta = \int h^*(\xi)\exp\{-(\xi_1^2+\ldots+\xi_k^2)\}\,\mathrm{d}\xi, \tag{3.6}$$
where $h^*(\xi)$ is a polynomial in the components of ξ.

The class of functions which can be well approximated by polynomial × normal forms is, in fact, rather rich and covers many of the $S_I[q(\theta)]$ integrands we typically encounter, provided the individual parameter components have support $(-\infty, \infty)$. In many problems this is, of course, not the case since the likelihood involves variance components or proportions defined in the interval $(0, 1)$, or whatever. However, the range of application of the normal × polynomial approximation is greatly extended by, in such cases, redefining the likelihood and prior in terms of transformed individual components: for example, by working with the logarithms of variance components and the logits of proportions. Visual evidence of the effectiveness of such transformations is provided in Smith *et al.* (1985, 1987), and such individual parameter transformations are a key feature of the strategies to be described later.

(b) *Parametrization issues*

A diagnostic plot for assessing approximate marginal posterior normality and suggesting effective reparametrization can be developed as follows. Let $p(\theta\,|\,x)$ denote the joint posterior density, with a mode at $\hat{\theta} = (\hat{\theta}_1, \ldots, \hat{\theta}_p)$, and let $p_j(\theta_j\,|\,x)$ denote the profile posterior for θ_j, defined by maximizing $p(\theta\,|\,x)$ over $\theta_i, i \neq j$, for each θ_j. A diagnostic proposed by Hills & Smith (1991*a*, *b*) is then to plot $T(\theta_j)$ against θ_j, where
$$T(\theta_j) = \mathrm{sgn}\,(\theta_j - \hat{\theta}_j)\{-2[\ln p_j(\theta_j\,|\,x) - \ln p(\hat{\theta}\,|\,x)]\}^{\frac{1}{2}}.$$
To motivate this procedure, consider the case of $p = 1$. If $p(\theta\,|\,x)$ is actually normal, we have
$$p(\theta\,|\,x) = p(\hat{\theta}\,|\,x)\exp\{-(1/2\sigma^2)(\theta-\hat{\theta})^2\},$$
for some σ^2, from which it follows immediately that $T(\theta)$ is linear in θ. In general, a Taylor expansion gives
$$\ln p(\theta\,|\,x) = \ln p(\hat{\theta}\,|\,x) + \tfrac{1}{2}H(\theta-\hat{\theta})^2 + O[(\theta-\hat{\theta})^3],$$
where H is the second derivative of $\ln p(\theta\,|\,x)$ evaluated at $\hat{\theta}$. It follows that
$$T(\theta) = (\theta-\hat{\theta})[-H - 2O[(\theta-\hat{\theta})^3]/(\theta-\hat{\theta})^2]^{\frac{1}{2}}.$$
The first term in the square root expression is observed Fisher information (and does not depend on θ). If this term is large, it will dominate and $T(\theta)$ will be approximately linear in θ (the 'large sample' case). Departure from linearity will then indicate the importance of the cubic term.

This '$T(\theta)$ against θ' diagnostic plot (which we shall refer to as the Bayes t-plot) is discussed in detail in Hills & Smith (1991, 1992) where the behaviour of the diagnostic plot is studied for various stylized cases. As an example, suppose that in fact $\ln(\theta - c)$ were distributed as $N(\mu, \sigma^2)$. In this case, it can be shown that
$$T(\theta) = \sigma^{-1}[\ln(\theta-c) - (\mu - \sigma^2)],$$
which has an asymptote at $\theta = c$.

Based on this and other theoretical forms of $T(\theta)$ against θ, 'look-up' rules can be established (see Hills & Smith (1991) for detailed derivations and illustration). In practice, accuracy of choices of constants (such as c in the above) is not critical. In any case, a second diagnostic check can be carried out for the suggested reparametrization.

More specific forms of reparametrization strategy may be suggested by particular statistical model classes. Consider, for example the autoregressive moving average processes, an ARMA (p,q) being specified by $\phi(B)\,y_t = \theta(B)\,\epsilon_t$, where y_t denotes an observation at time t, ϵ_t denotes white noise, B is the backward shift operator, $B^k y_t = y_{t-k}$, $\phi(B) = 1-\phi_1 B-\phi_2 B^2-\ldots-\phi_p B^p$ and $\theta(B) = 1-\theta_1 B-\theta_2 B^2-\ldots-\theta_q B^q$. For stationarity, the roots of $\phi(B) = 0$ must lie outside the unit circle; for invertibility (to ensure a unique model corresponding to the likelihood) the roots of $\theta(B) = 0$ must lie outside the unit circle. These conditions induce a constraint region, $C_p \times C_q$, for the model parameters (ϕ, θ), where $\phi = (\phi_1, \ldots, \phi_p)$, $\theta = (\theta_1, \ldots, \theta_q)$.

To motivate a suitable reparametrization in this case, first recall that a polynomial in x, with real coefficients,

$$1-\tau_1 x-\tau_2 x^2-\ldots-\tau_m x^m,$$

can be factorized as

$$\prod_{j=1}^{\frac{1}{2}m} (1-a_{1j}x-a_{2j}x^2) \quad \text{for} \quad m \text{ even},$$

$$(1-a_0 x) \prod_{j=1}^{\frac{1}{2}(m-1)} (1-a_{1j}x-a_{2j}x^2) \quad \text{for} \quad m \text{ odd},$$

where the a are also real. When the polynomial corresponds to $\phi(B)$ or $\theta(B)$, the condition that (ϕ, θ) lies in $C_p \times C_q$ is satisfied by ensuring that each point (a_{1j}, a_{2j}) lies in the triangle $|a_{2j}| < 1$, $|a_{1j}| < 1-a_{2j}$, with $|a_0| < 1$ if p or q is odd. A suitable reparametrization is then achieved by setting

$$a_{1j}^* = \ln\left(\frac{1-a_{2j}+a_{1j}}{1-a_{2j}-a_{1j}}\right), \quad a_{2j}^* = \ln\left(\frac{1+a_{2j}}{1-a_{2j}}\right),$$

with $a_0^* = \ln[(1+a_0)/(1-a_0)]$ if p or q is odd. For further details (and discussion of the many-to-one problem in passing from the τ to the a) see Marriott & Smith (1991).

After performing individual component transformations of the kind discussed above, the remaining problem of carrying out the linear transformations described in §3*a* is that the β_{ij}, μ_i and σ_i are not known. The strategies now to be described approach this problem by a form of statistical sequential learning process.

(*c*) *An iterative product rule strategy*

The polynomial $h^*(\xi)$, assumed to be of order d, say, appearing in the standardized form (3.6) can always be written as a linear combination of monomials of the form

$$\xi_1^{\alpha_1} \ldots \xi_k^{\alpha_k},$$

where $\alpha_1+\ldots+\alpha_k \leqslant d$ and $\alpha_i \geqslant 0$ for $i = 1, \ldots, k$. The required integral is thus a linear combination of products of the form

$$\prod_{i=1}^{k} \int_{-\infty}^{\infty} \xi_i^{\alpha_i} \exp(-\xi_i^2)\, d\xi_i.$$

The component integrals in this product are well-approximated by classical Gauss–Hermite quadrature rules, which have the general form

$$\int_{-\infty}^{\infty} f(t) \exp(-t^2)\, \mathrm{d}t \approx \sum_{i=1}^{n} w_i f(t_i),$$

where

$$w_i = 2^{n-1} n! \sqrt{\pi}/\{n^2[H_{n-1}(t_i)]^2\},$$

and t_i is the ith zero of the Hermite polynomial $H_n(t)$ (see, for example, Davis & Rabinowitz 1984).

This motivates the approximation of the original integral (3.2) by a cartesian product rule, which takes a weighted average of the integrand values at the intersection points of the k-dimensional grid corresponding to the Hermite zeros in each direction, with weights equal to the product of the corresponding weights. The numbers of zeros used (i.e. the choices of n in the Gauss–Hermite rule) can of course be different in the different directions. As a result of the polynomial × normal underlying assumption, following individual parameter component transformation, and provided the grids are chosen large enough, the same zeros and weights will serve to calculate normalizing constants, marginal density values and moments. This is a key feature of the approach, leading to considerable gains in efficiency.

However, as we remarked earlier, the β_{ij}, μ_i and σ_i implicit in the transformations motivating this approximation are unknown. We therefore proceed as follows, by an iterative, adaptive technique. We begin with small grids based on, say, four points in each direction, and initial estimates (possibly based on maximum likelihood) of the mean and covariance matrix of the posterior distribution (which imply estimates of β_{ij}, μ_i and σ_i). Based on these estimates, we perform cartesian product rule integrals, using the implied grid and weights, to obtain new estimates of the first and second moments of the posterior distribution and hence improved estimates of β_{ij}, μ_i and σ_i. This process is then iterated. The successive μ_i and σ_i control the centring and scaling of the grids; the β_{ij} control the orientation.

The general iterative Gauss–Hermite quadrature strategy may therefore be described as follows.

(i) Reparametrize individual parameters so that the resulting working parameters all take values on the real line.

(ii) Using initial estimates of the joint posterior mean vector and covariance matrix for the working parameters, transform further to a centred, scaled, more 'orthogonal' set of parameters.

(iii) Using the derived initial location and scale estimates for these 'orthogonal' parameters, perform cartesian product integration of functions of interest using suitably dimensioned grids.

(iv) Iterate, successively updating the mean and covariance estimates, until stable results are obtained both within and between grids of specified dimension.

The 'convergence' criterion implicit in (iv) above is, frankly, pragmatic in nature, and complete mathematical treatment of this iterative quadrature strategy seems very difficult. Since the algorithm is driven by up-dating of the first and second moment parameters of the gaussian kernel, the process can be viewed as a nonlinear iterative map on this moment space. Study of convergence thus reduces to the study of the behaviour of this nonlinear map. Some progress is reported in Shaw (1988*a*), who exhibits cases in which unique fixed points exist, but also cases with period cycling and even chaotic behaviour.

Table 1. *Comparison of points required for product and spherical rules*

parameter dimension k	degree 5 spherical (2^k+2k)	three-grid product 3^k	degree 7 spherical $(2^{k+1}+4k^2)$	four-grid product 4^k
3	14	27	52	64
4	24	81	96	256
5	42	243	164	1024
6	76	729	272	4096
7	142	2187	452	16384
8	272	6561	768	65536
9	530	19683	1348	262144

(d) *An iterative spherical rule strategy*

The problem with the product rule strategy outlined above is that above five or six dimensions it quickly becomes prohibitively expensive in terms of numbers of integrand calculations required, even for small grids: for example, a 4^6 grid requires 4096 evaluations and a 4^7 grid requires 16384 evaluations. In practice, we may need seven- or eight-point rules in many directions to achieve convergence, and this clearly precludes the routine use of product rules in high dimensions.

However, considerable gains in efficiency can be obtained by transforming to a spherical polar coordinate system and constructing optimal integration formulae based on symmetric configurations over concentric spheres. Such rules, discussed in detail in Stroud (1971, §§2.6, 2.7), are based on the observation that, if we make the transformation from ξ to ψ, where

$$\left.\begin{aligned} \xi_1 &= r\cos\psi_{k-1}\cos\psi_{k-2}\ldots\cos\psi_2\cos\psi_1,\\ \xi_2 &= r\cos\psi_{k-1}\cos\psi_{k-2}\ldots\cos\psi_2\sin\psi_1,\\ \xi_3 &= r\cos\psi_{k-1}\cos\psi_{k-2}\ldots\sin\psi_2,\\ &\ \ \vdots\\ \xi_k &= r\sin\psi_{k-1}, \end{aligned}\right\} \tag{3.7}$$

then the integral (3.6) becomes a product of integrals of the form

$$\int_{-\pi}^{\pi}(\cos\psi_j)^{a_j}(\sin\psi_j)^{b_j}\,d\psi_j,\quad j=1,\ldots,k-1,$$

and

$$\int_0^{\infty}(r^2)^{c-1}\exp(-r^2)\,dr,$$

for some a_j, b_j, c. The resulting optimal rules then take the form of a product of a Gauss–Laguerre rule for r^2, together with symmetric configurations on the spherical surfaces.

A rule which will integrate a (polynomial of degree d) × (multivariate normal) integrand exactly is called a 'degree d' rule. Table 1 compares the numbers of points required for optimal degree 5 and degree 7 product rules and spherical rules, respectively. The spherical rules are given in Stroud (1971); the product rules are the 3^k and 4^k Gauss–Hermite cartesian product rules.

It is clear from table 1 that spherical rules offer tremendous potential for efficient integration in the range of $4 \leqslant k \leqslant 9$, where the product grid strategy becomes too expensive.

(e) An iterative importance sampling strategy

The importance sampling approach to numerical integration is based on the observation that, if f and g are density functions,

$$\int f(x)\,\mathrm{d}x = \int [f(x)/g(x)]\,g(x)\,\mathrm{d}x = \int [f(x)/g(x)]\,\mathrm{d}G(x)$$
$$= E_G[f(X)/g(X)],$$

say, which suggests the 'statistical' approach of generating a sample from the distribution G and using the average of the values of the ratio f/g as an unbiased estimator of $\int f(x)\,\mathrm{d}x$. However, the variance of such an estimator clearly depends critically on the choice of G, it being desirable to choose g to be 'similar' to f.

In the univariate case, if we choose g to be heavier-tailed than f and if we work with $Y = g(X)$, the required integral is the expected value of $f[G^{-1}(X)]/g[G^{-1}(X)]$ with respect to a uniform distribution on the interval (0, 1). The resulting periodic nature of the ratio function over this interval then suggests that we are likely to get a reasonable approximation to the integral by generating 'uniformly distributed' random numbers. If f is a function of more than one argument (k, say), an exactly parallel argument suggests that the choice of a suitable g followed by the use of a suitably 'uniform' configuration of points in the k-dimensional unit hypercube will prove an acceptable alternative to the 'costly' procedure of generating 'random' uniformly distributed points in k-dimensions.

However, the effectiveness of all this depends on choosing a suitable G, bearing in mind that we need to have available a flexible set of possible distributional shapes, for which G^{-1} is available explicitly.

A suitable class of univariate distributions for importance sampling (due to Shaw 1988a) can be developed as follows.

Let U denote a uniform $U(0,1)$ random variable and let h denote a monotonic increasing function defined on the interval (0, 1), such that $h(u) \rightarrow -\infty$ as $u \rightarrow 0$. Now define a family of random variables by

$$X_A = Ah(u) + (1-A)\,h(1-u),$$

where $0 \leqslant A \leqslant 1$. Clearly, $A = \frac{1}{2}$ defines a random variable with a symmetric distribution; as $A \rightarrow 0$ or $A \rightarrow 1$ we obtain increasingly skewed distributions (in opposite directions). The tail behaviour of the distribution is governed by the choice of the function h.

Among interesting choices of the latter, we note: $h(u) = -[-\ln(u)]^k$, $k > 0$, which gives the logistic distribution for $k = 1$, $A = \frac{1}{2}$ and the exponential distribution for $k = 1$, $A = 0$; $h(u) = -\tan[\frac{1}{2}\pi(1-u)]$, whose symmetric member ($A = \frac{1}{2}$) is the Cauchy distribution; $h(u) = 1 - u^{-k}$, $k > 0$, which generalizes members of the Tukey-λ family of distributions.

If G_A, g_A denote, respectively, the distribution and density functions of X_A, we have:

$$G_A^{-1}(u) = x_A = Ah(u) + (1-A)\,h(1-u), \quad g_A(x) = G_A'(x).$$

If K_A, k_A denote the corresponding forms of G_A, g_A in terms of u, we have

$$K_A(u) = G_A(G_A^{-1}(u)) = [Ah'(u) + (1-A)\,h'(1-u)]^{-1}.$$

These forms guide the choices of h for which the corresponding importance sampling density is easy to use. Moreover, the moments of these families of distributions are polynomials in A (of corresponding order), the median is linear in A, and so on, so that sample information about such quantities provides (for any given choice of h) operational guidance about the appropriate choice of A.

In addition, the strategy requires the specification of 'uniform' configurations of points in the k-dimensional unit hypercube, a problem which has been extensively studied by number theorists. Systematic experimentation with various suggested forms of 'quasi-random' sequences has identified effective forms of configuration for importance sampling purposes. Based on measures of 'good lattice structure' motivated by the need for efficient calculation of (lower-dimensional) joint densities, Shaw (1988*b*) presents detailed comparison of a number of rational, irrational and irregular quasi-random sequence approaches to generating a configuration in the hypercube. For mathematical details, see Shaw (1988*b*).

If we combine the importance sampling and quasi-random ideas, the resulting general strategy is the following.

(i) Reparametrize individual parameters so that the resulting working parameters all take values on the real line.

(ii) Using initial estimates of the joint posterior mean vector and covariance matrix for the working parameters, transform further to a centred, scaled, more 'orthogonal' set of parameters.

(iii) In terms of these transformed parameters, set

$$g(x) = \prod_{j=1}^{k} g_j(x_j),$$

for 'suitable' choices of $g, j = 1, \ldots, k$.

(iv) use the inverse cumulative distribution function transformation to reduce the problem to that of calculating an average over a 'suitable' uniform configuration (quasi-random sequence) in the k-dimensional hypercube.

(v) Use information from this 'sample' to learn about skewness, tailweight, etc., for each $j = 1, \ldots, k$ and hence choose 'better' $g_j, j = 1, \ldots, k$, as well as revising estimates of the mean vector and covariance matrix.

(vi) Iterate until the sample variance of replicate estimates of the integral value is sufficiently small.

A variant of this strategy, which can be more effective if the standardized posterior density is approximately spherically symmetric, is to transform the quasi-random random configuration in the hypercube to a configuration on a spherical surface. See Shaw (1988*a*) for details.

In implementing these various quadrature and Monte Carlo techniques, it is often convenient to work with hybrid schemes. For example, combining a product rule for parameters of interest with a Monte Carlo rule for nuisance parameters, in order to facilitate eventual inference summaries, graphics, etc., for the parameters of interest.

Of course, all the above ideas can be combined with standard techniques of variance reduction, such as the use of antithetic variates: see, for example, Geweke (1988), with further ideas on useful importance sampling families illustrated in Geweke (1989).

4. Iterated sampling and resampling approaches

In the sequel, densities will be denoted, generically, by square brackets so that joint, conditional and marginal forms appear as $[X, Y]$, $[X \mid Y]$ and $[Y]$. The usual marginalization by integration procedure will be denoted by forms such as $[X] = \int [X \mid Y]*[Y]$.

(a) *Substitution sampling*

The substitution algorithm for finding fixed point solutions to certain classes of integral equations is a standard mathematical tool. Thus, for example, if X, Y, Z are random variables, so that, in the above notation

$$[X] = \int [X, Z \mid Y]*[Y], \quad [Y] = \int [X, Y \mid Z]*[Z], \quad [Z] = \int [Y, Z \mid X]*[X],$$

the marginal density $[X]$ of X is the fixed point solution of the equation

$$[X] = \int h(X, X')*[X'],$$

where the kernel is given by

$$h(X, X') = \int [X, Z \mid Y]*[X'', Y \mid Z']*[Y', Z' \mid X'],$$

a five-fold integral (with respect to X'', Y, Y', Z, Z').

The key idea of the stochastic substitution algorithm (see Tanner & Wong 1987; Gelfand & Smith 1990) is to estimate $[X]$ by successive stochastic simulation of random variates, drawn from the conditional distributions in the three above equations. The algorithm proceeds as follows: draw $X^{(0)}$ from an arbitrary $[X]_0$; draw $Y^{(0)'}, Z^{(0)'}$ from $[Y, Z \mid X^{(0)}]$; draw $X^{(0)'}, Y^{(1)}$ from $[X, Y \mid Z^{(0)'}]$; draw $X^{(1)}, Z^{(1)}$ from $[X, Z \mid Y^{(1)}]$ and then iterate. After t steps, with m replications of the process, we obtain $(X_j^{(t)}, Y_j^{(t)}, Z_j^{(t)})$, $j = 1, \ldots, m$. An estimate of the marginal density $[X]$ is then provided by

$$[\hat{X}]_t = \frac{1}{m} \sum_{j=1}^{m} [X \mid Y_j^{(t)}, Z_j^{(t)}].$$

In applications to bayesian inference, $[X, Y, Z]$ would denote the posterior distribution of unknown quantities of interest.

(b) *Gibbs sampling*

Throughout this section, we shall be dealing with collections of random variables $U_1, U_2, \ldots, U_k$, for which it is known that the joint density, $[U_1, U_2, \ldots, U_k]$, is uniquely determined by the full conditional densities $[U_s \mid U_r, r \neq s]$, $s = 1, 2, \ldots, k$. Our interest is typically in the marginal distributions, $[U_s]$, $s = 1, 2, \ldots, k$.

An algorithm for extracting marginal distributions from the full conditional distributions (in contrast to the form of conditionals used above in substitution sampling) was formally introduced as the Gibbs sampler in Geman & Geman (1984). The algorithm requires all the full conditional distributions to be 'available' for sampling, where 'available' is taken to mean that, for example, samples of U_s can be generated straightforwardly and efficiently from $[U_s \mid U_r, r \neq s]$, given specified values of the conditioning variables, $U_r, r \neq s$.

Gibbs sampling is a markovian updating scheme, which is a variant of the Metropolis algorithm. See, for example, Hastings (1970) and Peskun (1973) for seminal ideas on the use of Markov chain simulation algorithms for statistical problems. The Gibbs sampling algorithm proceeds as follows. Given an arbitrary starting set of values $U_1^{(0)}, \ldots, U_k^{(0)}$, we draw $U_1^{(1)}$ from $[U_1 \mid U_2^{(0)}, \ldots, U_k^{(0)}]$, then $U_2^{(1)}$ from $[U_2 \mid U_1^{(1)}, U_3^{(0)}, \ldots, U_k^{(0)}] \ldots$ and so on up to $U_k^{(1)}$ from $[U_k \mid U_1^{(1)}, \ldots, U_{k-1}^{(1)}]$ to complete one iteration of the scheme. After t such iterations we would arrive at $(U_1^{(t)}, \ldots, U_k^{(t)})$. Geman & Geman show under mild conditions that

$$U_s^{(t)} \xrightarrow{d} U_s \sim [U_s] \text{ as } t \to \infty.$$

Thus, for t large enough we can regard $U_s^{(t)}$ as a simulated observation from $[U_s]$.

Replicating this process m times produces m independent and identically distributed k-tuples $(U_{1j}^{(t)}, \ldots, U_{kj}^{(t)}), j = 1, \ldots, m$. For any s, the collection $U_{s1}^{(t)}, \ldots, U_{sm}^{(t)}$ can be viewed as a simulated sample from $[U_s]$. The marginal density could then be estimated by the finite mixture density.

$$[\hat{U}_s] = m^{-1} \sum_{j=1}^{m} [U_s \mid U_r = U_{rj}^{(t)}, r \neq s]. \tag{4.1}$$

(See Gelfand & Smith (1990) and Gelfand *et al.* (1990) for further discussion.)

Suppose interest centres on the marginal distribution for a variable V which is a function $g(U_1, \ldots, U_k)$ of $U_1, \ldots, U_k$. We note that evaluation of g at each of the $(U_{1j}^{(t)}, \ldots, U_{kj}^{(t)})$ provides samples of V, so that an ordinary kernel density estimate can readily be calculated.

In the bayesian framework, where U_s are unobservable, representing either parameters or missing data (and V can thus be a function of the parameters in which we are interested), all distributions will be viewed as conditional on the observed data, whence marginal distributions become the marginal posteriors needed for bayesian inference or prediction.

In many applications involving exponential families, some or all of the full conditionals may be familiar density forms from which sampling is straightforward. However, while 'conjugacy' simplifies the implementation of the Gibbs sampler it is not an essential element. In *any* Bayes model the full conditional distribution of any parameter is always identifiable from the joint density of the data and the parameters modulo normalizing constant. Using more sophisticated random variate generation approaches, such as the ratio of uniforms method (Devroye 1986; Wakefield *et al.* 1992), or methods which exploit features like log-concavity (Gilks & Wild 1991; Dellaportas & Smith 1991), we can sample the arbitrary non-normalized densities, although, of course, fine tuning of the sampling methodology, including 'clever' reparametrization, may be required to avoid highly inefficient random variate generation.

If the Gibbs sampler is run in order to generate an 'as if' independent sample from the joint posterior distribution (rather than simply to form estimates by ergodic averaging), this can be attempted either by replicate independent 'short' runs of the process or by extracting multiple sample values from a single 'long' run of the process. Output series need to be monitored by a stopping rule (formal or informal) which decides when 'convergence' has been achieved. In the case of replicate runs, the required 'independent' sample from the posterior is then formed by the final generated values from the stopped series.

The number of iterations to achieve 'convergence' is clearly a function of starting values and the correlation structure of the stochastic process generated by the Gibbs sampler. To try to get some insight into the importance and effect of such correlation (and hence into the possible importance of reparametrization to remove it), let us consider in detail the simple case of two parameters, where the joint posterior is actually zero-mean, unit-variance bivariate normal with correlation ρ. If we initialize the process at θ_2^0, say, the conditional distributions which drive the Gibbs sampler are given by

$$p(\theta_1^t \mid \theta_2^{t-1}) \equiv N(\rho\theta_2^{t-1}, 1-\rho^2), \quad p(\theta_2^t \mid \theta_1^t) \equiv N(\rho\theta_1^t, 1-\rho^2),$$

from which it follows straightforwardly that the joint distribution of (θ_1^t, θ_2^t) has means $\rho^{2t-1}\theta_2^0, \rho^{2t}\theta_2^0$, variances $1-\rho^{4t-2}, 1-\rho^{4t}$ and correlation $\rho[(1-\rho^{4t})/(1-\rho^{4t-2})]^{\frac{1}{2}}$, so that the effects of θ_2^0 and ρ on the rate of convergence can easily be examined.

The clear (and intuitive) message is that really high correlations disastrously slow down the convergence of the Gibbs sampler and that the higher the correlation the more serious are bad starting values. At the other end of the scale, small correlations (even of the order of $\rho = 0.8$) imply relatively trivial numbers of iterations to convergence in this simple case, with bad starting values quickly forgotten. Breaking high correlation can be achieved by means of an application after a few initial iterations (say, 10–15) of the orthogonalizing transformation defined earlier in §3*c*.

Complete implementation of the Gibbs sampler requires that a determination of t be made and that, across iterations, choice(s) of m specified. See Gelfand *et al.* (1990) for further discussion of convergence issues. See also, Gelfand & Smith (1990) and Carlin *et al.* (1991), as well as a discussion of other markovian updating procedures by Aykroyd & Green (1991).

(c) *Resampling*

As a first step towards motivating the resampling approach, we note the essential duality between a sample and the density (distribution) from which it is generated. Clearly, the density generates the sample; conversely, given a sample we can approximately recreate the density (as a histogram, an empirical cumulative distribution function or whatever).

Suppose we now shift the focus in (1.1) from densities to samples. In terms of densities, the inference process is encapsulated in the updating of the prior density, $p(\theta)$, to the posterior density, $p(\theta \mid x)$, through the medium of the likelihood function, $l(\theta; x)$. Shifting to samples, this corresponds to the updating of a sample from $p(\theta)$ to a sample from $p(\theta \mid x)$ through the likelihood function $l(\theta; x)$.

Generally, suppose that a sample of random variates is easily generated, or has already been generated, from a continuous density $g(\theta)$, but that what is really required is a sample from a density $h(\theta)$ absolutely continuous with respect to $g(\theta)$. Can we somehow utilize the sample from $g(\theta)$ to form a sample from $h(\theta)$? Slightly more generally, given a positive function $f(\theta)$ which is normalizable to such a density $h(\theta) = f(\theta)/\int f(\theta)\,d\theta$, can we form a sample from the latter given only a sample from $g(\theta)$ and the functional form of $f(\theta)$?

We may approximately resample from $h(\theta) = f(\theta)/\int f(\theta)\,d\theta$ as follows. Given θ_i, $i = 1, \ldots, n$, a sample from g, calculate $\omega_i = f(\theta_i)/g(\theta_i)$ and then

$$q_i = \omega_i / \sum_{j=1}^{n} \omega_j.$$

Draw θ^* from the discrete distribution over $\{\theta_1, \ldots, \theta_n\}$ placing mass q_i on θ_i. Then θ^*

is approximately distributed according to h with the approximation 'improving' as n increases (see Smith & Gelfand 1992). Note that this procedure is a variant of the by now familiar bootstrap resampling procedure. The usual bootstrap provides equally likely resampling of the θ_i, while here we have weighted resampling with weights determined by the ratio of f to g. See, also, Rubin (1988), who refers to this procedure as SIR (sampling/importance resampling).

Several obvious uses of this sampling–resampling perspective are immediate. In general, the translation from functions to samples provides a wealth of opportunities for creative exploration of bayesian ideas and calculations in the setting of computer graphical and exploratory data analysis tools. Also, we can easily approach problems of sensitivity of inferences to model specification, such as: How does the posterior change if we change the prior? How does the posterior change if we change the likelihood?

In the density function/numerical integration setting, such sensitivity studies are rather off-putting, in that each change of a functional input typically requires one to carry out new calculations from scratch. This is not the case with the sampling–resampling approach, as we now illustrate in relation to the questions posed above.

In comparing two models in relation to the second question, we note that change in likelihood may arise in terms of (i) change in distributional specification with θ retaining the same interpretation, e.g. a location; (ii) change in data to a larger data-set (prediction), a smaller data set (diagnostics), or a different data-set (validation).

To unify notation, we shall in either case denote two likelihoods by $l_1(\theta)$ and $l_2(\theta)$. We denote two different priors to be compared in relation to the first question by $p_1(\theta)$ and $p_2(\theta)$. For complete generality, we consider changes to both l and p, although in any particular application we would not typically change both. Denoting the corresponding posterior densities by $\tilde{p}_1(\theta), \tilde{p}_2(\theta)$, we easily see that

$$\tilde{p}_2(\theta) \propto [l_2(\theta)\, p_2(\theta)/l_1(\theta)\, p_1(\theta)]\, \tilde{p}_1(\theta). \tag{4.2}$$

Letting $v(\theta) = l_2(\theta)\, p_2(\theta)/l_1(\theta) p_1(\theta)$, we may directly apply the weighted bootstrap method to (4.2) taking $g = \tilde{p}_1(\theta), f = v(\theta)\, \tilde{p}_1(\theta)$ and $\omega_i = v(\theta_i)$. Resampled θ^* will then be approximately distributed according to f standardized, which is precisely $\tilde{p}_2(\theta)$.

5. Overview

(*a*) *Analytic against numerical integration procedures*

Of the analytic approximation techniques available, those based on the Laplace approximation (§2*c*) seem to be the most systematically studied and validated. The choice between these and numerical integration procedures rests largely on the dimensionality and complexity of the problem. As the latter becomes greater, implementation of the analytic techniques becomes very difficult indeed.

(*b*) *Quadrature against Monte Carlo integration*

Here again, dimensionality and complexity of the posterior functional forms are the main determining factors. Roughly speaking, for relatively well-behaved functions (typically following reparametrization) product rules can be effective in up to about six dimensions, with spherical rules extending the domain of quadrature up to nine dimensions. However, beyond that (or even for lower-dimensional problems with badly behaved functions or awkward parameter constraints) Monte Carlo methods are generally necessary.

(c) *Importance sampling against iterative sampling*

Use of importance sampling for high-dimensional problems is something of an art form, both in the choice of effective importance sampling functions and in the design of variance reduction sampling strategies. However, if a good procedure can be found, it is likely to be computationally more efficient than a Markov chain based iterative sampling procedure. On the other hand, the latter, and, in particular, the Gibbs sampler, has the merit of being (typically) very easy to implement, requiring very little numerical or stochastic simulation expertise. Moreover, in very complex problems it may simply prove too difficult to identify any suitable importance sampling strategy.

To give a concrete sense of the different 'flavour' of the two approaches, we return briefly to the mixture analysis problem of §1*b*.

Shaw (1988*c*) provides a detailed analysis for a $k\,(=5)$ component univariate normal mixture model for fish lengths. Lengths are binned into n class ranges, resulting in N_j fish assigned to class j, corresponding to the interval $[x_j, x_{j+1})$, with $N_1+\ldots+N_n = N$. If (μ_i, σ_i, p_i) denote the k component means, variances and proportions, it is easy to see that the log-likelihood is given by

$$\sum_{j=1}^{n} N_j \ln \hat{f}_j - N \ln (\hat{F}_{n+1} - \hat{F}_1),$$

where

$$\hat{F}_j = \sum_{i=1}^{k} p_i\, \Phi[(x_j - \mu_i)/\sigma_i],$$

and $\hat{f}_j = \hat{F}_{j+1} - \hat{F}_j$ (with $\Phi(\cdot)$ denoting the standard normal cumulative distribution function). This log-likelihood is then combined with a prior specification for the $3k-1$ $(=14)$ unknown parameters, the prior chosen to reflect basic information about fish growth, variability and abundance in five successive cohorts. In particular, such knowledge constrains all parameters to be positive, the means to be increasing and the proportions to lie in the simplex. This is a case where parameter transformation is vital and Shaw worked with

$$\begin{aligned}
\theta_1 &= \mu_1 - 11 && \text{(for numerical stability)},\\
\theta_{3i-2} &= \ln(\mu_i - \mu_{i-1}) && (i = 2, \ldots, k),\\
\theta_{3i-1} &= \ln \sigma_i && (i = 1, \ldots, k),\\
\theta_{3i} &= \ln\left[p_i \Big/ \left(1 - \sum_{j=1}^{i} p_j\right)\right] && (i = 1, \ldots, k-1).
\end{aligned}$$

A detailed exploration and summary of the resulting 14-dimensional posterior density for the θs was accomplished with a quasi-random spherical rule (as described at the end of the §3), using 10000 nodes on two spherical shells. Further details are given in Shaw (1988*c*). The point to note in the context of the present discussion is that the posterior full conditionals for each of the 14 parameters (i.e. for each parameter given the values of the other 13) seem to be extremely messy forms, as a consequence of the complicated form of the log-likelihood, so that Gibbs sampling looks no easier than direct numerical integration.

Now consider the same k component normal model, but with precisely observed (rather than grouped) observations, so that the likelihood takes the form

$$\prod_{l=1}^{N} \sum_{i=1}^{k} p_i \phi[(x_l - \mu_i)/\sigma_i],$$

where $\phi(\cdot)$ denotes the standard normal PDF. A surprisingly simple structure for the Gibbs sampler can be obtained by introducing, as further unknowns, indicator quantities $Z_1, \ldots, Z_N$, such that $Z_l = i$ corresponds to observation x_l actually coming from component i. In this case, if (μ_i, σ_i) are assigned conjugate normal-inverse-gamma priors, $(p_1, \ldots, p_k)$ a Dirichlet prior and the Z_l uniform priors, it is easy to show that successive generation from the full conditionals reduces to: draw the Zs from a specified discrete distribution; the ps from a Dirichlet distribution; the μs from a normal distribution and the σs from an inverse gamma distribution. Substantial iterative computation is then required, but the need for sophisticated numerical understanding on the part of the statistical analyst is obviated.

Much of the author's work reviewed here was supported by the SERC's Complex Stochastic Systems Initiative.

References

Aykroyd, R. G. & Green, P. J. 1991 Global and local priors and the location of lesions using gamma camera imagery. *Phil. Trans. R. Soc. Lond.* A **337**, 323–342. (This volume.)

Carlin, B. P., Gelfand, A. E. & Smith, A. F. M. 1991 Hierarchical bayesian analysis of change-point problems. *Appl. Statist.* (In the press.)

Davis, P. J. & Rabinowitz, P. 1984 *Methods of numerical integration*, 2nd edn. Orlando, Florida: Academic Press.

Dellaportas, P. & Smith, A. F. M. 1991 Bayesian inference for generalized linear and proportional hazards models via Gibbs sampling. *Appl. Statist.* (In the press.)

Devroye, L. 1986 *Non-uniform random variate generation*. Springer-Verlag: New York.

Gelfand, A. E. & Smith, A. F. M. 1990 Sampling based approaches to calculating marginal densities. *J. Am. statist. Ass.* **85**, 398–409.

Gelfand, A. E., Hills, S. E., Racine-Poon, A. & Smith, A. F. M. 1990 Illustration of bayesian inference in normal data models using Gibbs sampling. *J. Am. statist. Ass.* **85**, 972–985.

Geman, S. & Geman, D. 1984 Stochastic relaxation, Gibbs distributions and the bayesian restoration of images. *IEEE Trans. Patt. Analysis Mach. Int.* **6**, 721–741.

Geweke, J. 1988 Antithetic acceleration of Monte Carlo integration in Bayesian inference. *J. Econometrics* **38**, 73–90.

Geweke, J. 1989 Bayesian inference in econometric models using Monte Carlo integration. *Econometrika* **57**, 1317–1339.

Gilks, W. R. & Wild, P. 1991 Adaptive rejection sampling for Gibbs sampling. *Appl. Statist.* (In the press.)

Hastings, W. K. 1970 Monte Carlo simulation methods using Markov chains and their applications. *Biometrika* **57**, 97–109.

Hills, S. E. & Smith, A. F. M. 1991 Diagnostic plots for improved parameterization in Bayesian inference. (Submitted.)

Hills, S. E. & Smith, A. F. M. 1992 Parameterization issues in Bayesian inference. In *Bayesian statistics 4* (ed. J. Bernardo *et al.*). Oxford University Press.

Kass, R. E., Tierney, L. & Kadane, J. B. 1989 Asymptotics in Bayesian calculation. In *Bayesian statistics 3* (ed. J. M. Bernardo *et al.*), pp. 261–278. Oxford University Press.

Leonard, T., Hsu, J. S. J. & Tsu, K.-W. 1989 Bayesian marginal inference. *J. Am. statist. Ass.* **84**, 1051–1058.

Lindley, D. V. 1980 Approximate Bayesian methods. In *Bayesian statistics* (ed. J. M. Bernardo *et al.*), pp. 223–245. Valencia University Press.

Marriott, J. M. & Smith, A. F. M. 1991 Reparameterization aspects of Bayesian methodology for ARMA models. *J. Time Series Analysis.* (In the press.)

Naylor, J. C. & Smith, A. F. M. 1982 Applications of a method for the efficient computation of posterior distributions. *Appl. Statist.* **31**, 214–225.

Peskun, P. H. 1973 Optimum Monte Carlo sampling using Markov chains. *Biometrika* **60**, 607–612.

Rubin, D. B. 1988 Using the SIR algorithm to simulate posterior distributions. In *Bayesian statistics 3* (ed. J. M. Bernardo *et al.*), pp. 395–402. Oxford University Press.

Shaw, J. E. H. 1988*a* Numerical and graphical techniques for Bayesian inference. Ph.D. thesis, University of Nottingham.

Shaw, J. E. H. 1988*b* A quasi-random approach to integration in bayesian statistics. *Ann. Statist.* **16**, 895–914.

Shaw, J. E. H. 1988*c* Aspects of numerical integration and summarization. In *Bayesian statistics* 3 (ed. J. M. Bernardo *et al.*), pp. 411–428. Oxford University Press.

Smith, A. F M. & Gelfand, A. E. 1992 Bayesian statistics without tears: a sampling-resampling perspective. *Am. Statist.* (In the press.)

Smith, A. F. M., Skene, A. M., Shaw, J. E. H. & Naylor, J. C. 1987 Progress with numerical and graphical methods for bayesian statistics. *Statistician* **36**, 75–82.

Smith, A. F. M., Skene, A. M., Shaw, J. E. H., Naylor, J. C. & Dransfield, M. 1985 The implementation of the Bayesian paradigm. *Commun. Statist.* A**14**, 1079–1102.

Stroud, A. H. 1971 *Approximate calculation of multiple integrals.* New Jersey: Prentice-Hall.

Tanner, M. & Wong, W. 1987 The calculation of posterior distributions by data augmentation. *J. Am. statist. Ass.* **82**, 528–550.

Tierney, L. & Kadane, J. B. 1986 Accurate approximation for posterior moments and marginal densities. *J. Am. statist. Ass.* **81**, 82–86.

Tierney, L., Kass, R. E. & Kadane, J. B. 1989 Approximate marginal densities of nonlinear functions. *Biometrika* **76**, 425–434.

Titterington, D. M., Smith, A. F. M. & Makov, U. E. 1986 *Statistical analysis of finite mixture distributions.* Chichester: Wiley.

Wakefield, J. C., Gelfand, A. E. & Smith, A. F. M. 1992 Efficient generation of random variates via the ratio-of-uniforms method. *Statist. Computing.* (In the press.)

Probabilistic expert systems and graphical modelling: a case study in drug safety

By David J. Spiegelhalter[1], A. Philip Dawid[2], Tom A. Hutchinson[3] and Robert G. Cowell[2]

[1] *MRC Biostatistics Unit, 5 Shaftesbury Road, Cambridge CB2 2BW, U.K.*
[2] *Department of Statistical Science, University College London, U.K.*
[3] *Royal Victoria Hospital, Montreal, Canada*

Probabilistic expert systems are intended to provide reasoned guidance in complex environments characterized by extensive uncertainty. An explicit 'causal' model is constructed for the process being observed, in which an acyclic directed graph is used to express conditional independence assumptions about variables, and probability assessments specify a full joint probability distribution. The resulting graphical structure can cope with a range of issues that arise in realistic modelling. Here we consider a particular example of assessing the chance that a suspected adverse reaction is due to a particular drug under suspicion. The background biological knowledge provides an appropriate model and probability assessments are obtained from expert microbiologists. The model allows a variety of interpretations for 'causality'. Details of the graphical and computational algorithms used to perform efficient calculations of conditional probabilities on complex graphical structures are provided and illustrated with the example. Further developments should allow updating of the risk parameters in the light of a series of case reports, and may form the basis for a flexible expert system for causality assessment and post-marketing surveillance.

1. Introduction

Expert systems are computer programs that are intended to provide reasoned guidance in complex situations. Uncertainty pervades many applications, due to missing information or an inability to build a purely logical model for the process being studied, and a variety of approaches have been suggested for dealing with this problem: fuzzy logic, certainty factors, non-monotonic reasoning, belief functions have all been suggested; see Shafer & Pearl (1990) for a wide collection of papers. The probabilistic approach has been hampered by apparent computational difficulties, but recent developments have now made it feasible to deal with real applications using specialist software for probabilistic reasoning.

In this paper we select a particular example to illustrate the methodology, which emphasizes the use of background scientific knowledge as a basis for a model that can be used for handling individual cases in a coherent way. The problem is in the context of routine post-marketing surveillance of pharmaceutical products, which requires the processing of reports of suspected adverse reactions. The accurate identification of the cause of the reaction is of great importance both to the pharmaceutical industry and to regulatory authorities, and yet the poor quality of the received data can make such 'causality assessment' exceptionally difficult.

Phil. Trans. R. Soc. Lond. A (1991) **337**, 387–405
Printed in Great Britain

In the next section we briefly review current methods of causality assessment, including a probabilistic procedure, and in §3 we introduce a specific problem concerning the development of a diarrhoeal reaction (pseudomembranous colitis (PMC)) following antibiotic therapy. In §§4 and 5 we step through the stages in representing the PMC problem as a 'causal network', showing how the basic structure can be extended in a modular fashion to provide an increasingly realistic and complex model. Our description emphasizes the exploratory nature of our model building but, while certainly not claiming to have a definitive representation of the process under study, we hope to demonstrate the concise manner in which complex phenomena can, with care, be modelled by this approach. In §6 we summarize technical aspects of handling causal network representations of disease processes that have been developed in the context of expert systems, emphasizing the intuitive graphical structure and its re-representation for computational purposes.

We then give an example based on an actual case report, and demonstrate sensitivity analysis of the conclusions to assumptions about the context in which the reaction developed.

The conclusions that the model draws about the drug responsible for a reaction in a specific patient depend crucially on the assumptions concerning the potential of that drug to initiate the chain of events that led to the reaction, and these assumptions are open to question. Thus, while identifying the cause in individual cases is certainly important, it is even more vital to use the reports to learn more about the propensity of a drug to cause harm. This inverse process is difficult in view of the mainly highly selected nature of the reports that are received, but some tentative suggestions are made in our concluding section.

The general framework of our approach is currently being explored in a variety of applications, which include the diagnosis of congenital heart disease (Spiegelhalter & Cowell 1991), monitoring in anaesthesia (Beinlich *et al.* 1989), diagnosis of neurological disorders (Andreassen *et al.* 1989) and decision making in diabetes (Andreassen *et al.* 1991). Each of these applications requires probabilistic identification of an underlying causal mechanism that will best explain the observed constellation of findings on individual cases; they are not concerned with general identification of causal mechanisms for populations. This paper emphasizes the technical issues: Hutchinson *et al.* (1991*b*) and Cowell *et al.* (1991), respectively, discuss our problem from the medical and computing points of view.

2. Causality assessment: background and an example

The drug surveillance group of a pharmaceutical company or regulatory authority will regularly receive reports of adverse clinical events that are suspected to be reactions to previous therapy. The reports will usually give some background information on the patient, the date and type of reaction, the dates on which the suspected drug was taken, and possibly the effects of stopping the drug ('dechallenge'). However, vital information is often missing, such as other drugs being taken and the clinical context in which the reaction occurred.

Four basic methods for processing these reports can be identified (Hutchinson & Lane 1989). First, the global approach requires the assessor to examine the evidence and make an intuitive overall judgement as to whether the reaction was certainly/probably/possibly/unlikely to have been caused by the drug in question. The irreproducibility of such an approach has been shown in observer agreement studies (Karch *et al.* 1976), and this has led to two more structured procedures. The algorithmic, or flowchart, method requires the assessor to step through a sequence of questions that logically lead to a conclusion. For example, a U.S. Food and Drug Administration algorithm (Turner 1984) first asked if the current has a reasonable temporal association from the drug, and if so it asked if there was a dechallenge from the drug (i.e. did the reaction cease when the drug was stopped); if this information is not available, the algorithm immediately classified the causal relationship as 'possible' and asked no more questions.

Such algorithms seem unnecessarily rigid and make limited use of available information; this has led to scoring systems in which points are accumulated from aspects of the case evidence, and if the total points lie in a certain interval then the assessment is 'probable', etc. (Venulet 1984). Hutchinson & Lane (1989) have pointed out that such scores are arbitrary and ignore the relative strengths of the items of evidence, and with other authors (Lane *et al.* 1987; Lane 1989) they have suggested a bayesian probabilistic approach, in which the probability of causation is explicitly calculated. We emphasize that, in the following development, we do not claim that the idea of 'causation' can be uniquely defined. A variety of interpretations are possible, but we shall try to be precise as to which we are using in any context. As we shall find in §5, one of the major strengths of the bayesian graphical approach is that it allows a number of quantities to be calculated, each of which could be argued to be the probability of 'causation'.

Suppose there are two competing causes, A and B, of a specified type of reaction E, where B could represent 'natural causes'. Let $A \rightarrow E$ denote the event that E occurred and that A alone 'caused' E, which we take to mean that if A alone had been given then E would still have occurred exactly as it did (in the following sections we consider assigning joint responsibility to two drugs simultaneously). Let the background information (type of patient, drug dosage and schedule, class of reaction, time horizon within which the occurrence of the specific clinical problem is considered as a reaction) be denoted by M. Then the odds on A against B causing the reaction, before incorporating the occurrence of specific details of the reaction, are

$$p(A \rightarrow E \mid M)/p(B \rightarrow E \mid M). \tag{1}$$

After detailed findings F are obtained (e.g. timing of the reaction relative to time of taking the drugs, effects of dechallenge, rechallenge, etc.), the revised posterior odds are, by Bayes's theorem,

$$\frac{p(A \rightarrow E \mid F, M)}{p(B \rightarrow E \mid F, M)} = \frac{p(F \mid A \rightarrow E, M)}{p(F \mid B \rightarrow E, M)} \times \frac{p(A \rightarrow E \mid M)}{p(B \rightarrow E \mid M)}, \tag{2}$$

or

$$\text{posterior odds} = \text{relative likelihood} \times \text{prior odds}.$$

The relative likelihood is an assessment of how much more likely is the specific case evidence were A to have caused the reaction as against B. Figure 1 displays a 'high-level' representation of the process being modelled.

The arrows indicate dependencies to be taken into account when considering the probability of any random (elliptical) node in such a graph (rectangular nodes indicate fixed context). Thus figure 1 represents terms in expressions (1) and (2): the probability of each possible cause depends directly on case parameters (M), while the probability of case findings (F) depends on both cause and background parameters. The crucial element is that while a graph may be constructed by thinking causally,

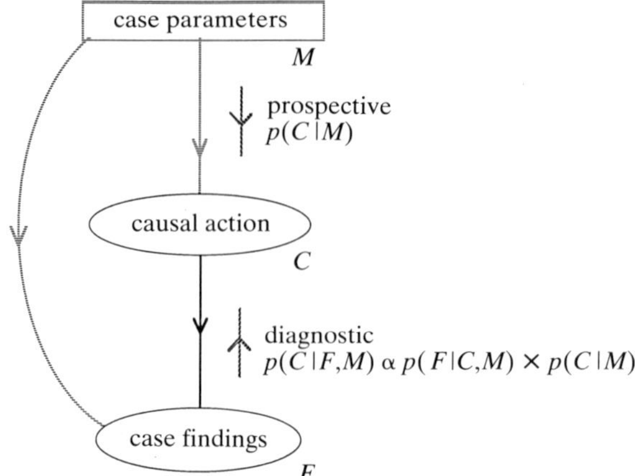

Figure 1. Overview of process being modelled: prospective modelling of causal mechanism given case parameters, and specific case findings given case parameters and cause. After observing F, Bayes's theorem 'reverses the arrow' between C and F, and allows the calculation of the posterior probability of cause given M and F.

Table 1. *Background and reaction details of patient (timings relative to start of drug A)*

background information M	
patient	female, aged 29
drugs given	antibiotic A day 0–day 10
	antibiotic B day 12–day 22
reaction observed	pseudomembranous colitis (PMC)
details (F) of reaction	
timing	occurred on day 27

when specific case findings are actually observed this diagnostic evidence needs to be propagated back to the causal action node. Figuratively the arrow from causal action to case findings needs to be reversed; technically this is precisely the role of the likelihood terms used in Bayes's theorem (2).

In many examples it may be reasonable to assume the items of evidence in D are conditionally independent given the true cause, and then the relative likelihood breaks into a product of terms that may be individually assessed. Examples of such analyses are given in the Proceedings of a Drug Information Workshop (1986). However, we now introduce an example which appears to demand additional sophistication in its probabilistic modelling.

3. A specific example of a reaction

We now consider an actual case report received by the drug surveillance department of a pharmaceutical company. The relevant background and details of the reaction are given in table 1.

In a typical scoring system each drug is assessed individually without directly comparing their likelihoods of having led to the event. In contrast, a probabilistic approach as described in the previous section would first assess the prior odds that A rather than B would cause a reaction within some time horizon that defines the occurrence of a reaction, and then assess the relative likelihood of the reaction

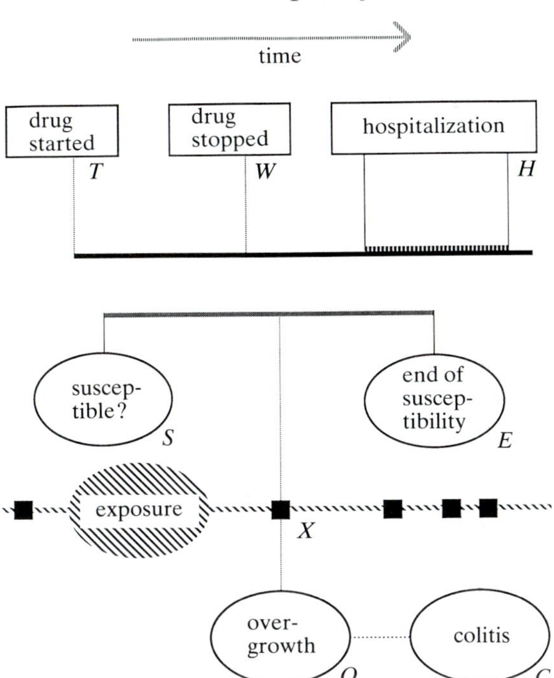

Figure 2. Schematic diagram for process under study. The top level represents the timing of the treatment of the patient, and the possible 'window of susceptibility' is shown below. Next comes the exposure process to *C. difficile*, which is shown as being intensified while the patient is in hospital. If exposure occurs within the window of susceptibility then overgrowth is assumed to happen, followed a variable number of days later by clinical evidence of colitis.

occurring on day 27 (the only detailed information), given it occurred within the specified horizon. See Hutchinson *et al.* (1991*a*) for an example of such an analysis using a spreadsheet program.

Such an assessment is, however, somewhat unsatisfactory since a considerable amount is known about the mechanism by which an antibiotic can lead to the development of PMC, and a conscientious assessment should take this knowledge into account, explicitly and quantitatively. In collaboration with an expert microbiologist, Dr Douglas Burdon, we have developed a formal model to describe this process. Specifically, an antibiotic depletes the bowel flora in such a way that, if the patient is then exposed to the particular virulent bacterial organism *Clostridium difficile* (abbreviated to *C. difficile*), this organism may colonize the gut. It is this overgrowth that after a few days will lead to clinical signs of PMC-induced diarrhoea.

The process is shown in figure 2. A drug is started at time T (taken as 0 in the computer analysis) and taken until time W. Soon after T there is a chance that a 'window of susceptibility' is opened up due to modification of gut flora (time S), and this stays open until time E, which is some unknown period after the drug is stopped. In all our analysis we have assumed that $S = T+1$ if the patient becomes susceptible; i.e. if the drug is going to alter the gut flora, it does so by the end of the first day of treatment. Meanwhile, and quite independently, there is an 'exposure process' going on in which each day there is a certain chance of being exposed to *C. difficile*. The day of first exposure after time $T+1$ is the 'relevant' exposure day (X). If this occurs within the window of susceptibility, then on this day overgrowth occurs

(O), and soon after colitis is observed (C). The diagram shows that if the patient is in hospital, particularly in a ward where there is an epidemic of *C. difficile*, then the exposure rates may increase dramatically.

Various additional complexities may be relevant. First, after two drugs have been given, then even if they act independently both may have their windows of susceptibility open when an exposure occurs and so the unique cause is not ascribable, i.e. if either drug had not been given the reaction would still have occurred, but equally if either drug had been given alone the reaction would have occurred. Secondly, hospitalization and the existence of an epidemic may change the daily exposure rate. Thirdly, it is possible that certain antibiotics, particularly when given intravenously, may actually 'block' overgrowth while they are being administered since they may also act on *C. difficile*, and hence relevant exposure can only take place after the drug has been stopped. Finally, most of the relationships described in the model are stochastic and their parameters can only be assessed subjectively by expert microbiologists on the basis of limited experience: examples include the daily chance of exposure of *C. difficile* in the community, the extent to which susceptibility continues after the drug is stopped, the chance an individual becomes susceptible at all and the lag between overgrowth and development of colitis.

The complexity of the problem strongly suggests that intuitive assessments of likelihood ratios may not be reliable, and hence we have investigated a formal analysis using a graphical approach that uses recent developments within the general context of expert systems.

4. A bayesian graphical representation of the causality assessment problem

Systems intended to help clinicians classify patients into diagnostic or prognostic states exist in similar degrees of development to those in causality assessment. There are global judgments in which only structured information is available, algorithms, developed either informally or based on data analysis, and scoring systems with their weights again either assessed informally or through statistical discriminant analysis. Recent work in artificial intelligence in medicine has, however, emphasized the use of causal networks, in which an attempt is made explicitly to model the qualitative medical knowledge that is available concerning the disease processes under consideration (Szolovits *et al.* 1988). These models are represented by directed graphs in which links represent direct influences, and absent links indicate conditional independence assumptions. Attached to these links are conditional probability tables that quantitatively express the uncertainty on the links, and the aim is to be able to observe fragmentary evidence, possibly sequentially, on parts of the graph, and propagate the effect of this evidence through the graph to revise the probability of the nodes of direct interest. Such graphical representations have a long history, beginning with Wright (1934) and continuing in the areas of structural modelling in the social sciences, while there is also a clear relationship with the pedigree graphs used in genetics (Thompson 1986).

In our context we need to develop a directed graphical representation for the qualitative mechanism by which a reaction might occur when one or two antibiotics had been previously taken, and provide assessments of the necessary conditional probability distributions. The theory for probability calculations on such networks has been developed by Pearl (1986, 1988) and Lauritzen & Spiegelhalter (1988), and in §6 we describe the computational algorithm by which these initializing conditional probabilities may be manipulated to allow efficient calculation of the required outputs once evidence on a case has been obtained; the required outputs will generally be the probabilities of causation and the evidence will generally be the day on which colitis is noted, although the computational procedures are not restricted to these circumstances.

There are four main stages in constructing a network and turning it into a representation ready to receive evidence, and two further stages in processing evidence as it arrives.

(*a*) *Representation of the qualitative knowledge as a directed acyclic graph* (DAG)

The nodes of the graph are random variables and direct influences of a node v are 'parents' of v (denoted $\mathrm{pa}(v)$) in the graph. Formally, the graph represents the assumption that the joint distribution of the whole set V can be expressed as

$$p(V) = \Pi_v\, p(v \mid \mathrm{pa}(v)), \tag{3}$$

i.e. the joint distribution is made up of terms which model the conditional distribution of each node given its direct influences. This is equivalent to a conditional independence assumption that, if we are told the values of $\mathrm{pa}(v)$, then v is independent of all other nodes in the graph except direct descendants of v (see Lauritzen *et al.* (1990) for a full description of the conditional independence statements that can be read off a DAG). Such assumptions follow naturally in a context such as genetics (Thompson 1986), in which a child's genotype is only directly influenced by its parents, and so if the parents' genotypes are known then no other individual in a pedigree provides information about the child's genotype (apart from descendants of the child). The essence of our approach is to make similar conditional independence assumptions within other domains, and then exploit (3) to allow the full joint distribution to be obtained by a set of assessments of individual terms $p(v \mid \mathrm{pa}(v))$.

We first consider the simplified situation in which only one drug has been taken; this is appropriate if we were to assume either A or B was the cause of the reaction, with no allowance made for simultaneous causation. Thus in each one-drug model we are assuming we know the causative agent, but we may use the model to calculate prior and likelihood terms for A and B individually, to be processed using (2).

Figure 3 shows a network representation of the essential elements of the complex processes summarized in figure 2. Each elliptical node represents a random quantity, specifically the time, including ∞ to represent 'never' where appropriate, of the critical events defined in the legend. The boxes represent fixed contexts defined by the background information on the case. The 'expert' biological knowledge allows us to make the conditional independence assumptions represented by the graph; for example, that given the day on which overgrowth occurs, the day on which colitis first appears is independent of all other events in the network. Each such assumption translates into a formal conditional probability statement, for example, $p(C \mid O, \text{other events}) = p(C \mid O)$, and we obtain from (3) an expression for the joint distribution of all the unknown quantities:

$$p(C, O, E, S, X \mid M = \{H, T, W\}) = p(C \mid O)\, p(O \mid E, S, X)\, p(E \mid S, W)\, p(S \mid T)\, p(X \mid H), \tag{4}$$

for which we need only obtain assessments of the distribution of each quantity conditional on its direct influences (parents) in the graph. Once we have such

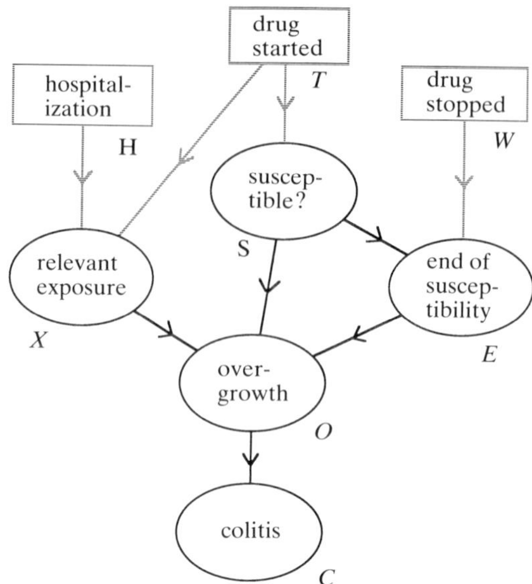

Figure 3. Graphical model of process when only one drug is being taken. Case parameters: H, days of entering and leaving hospital; T, day treatment started; W, day treatment withdrawn. Internal variables: X, day of first relevant exposure to *C. difficile*; S, day of start of susceptibility to overgrowth; E, day susceptibility ends; O, day overgrowth occurs. Case findings: C, day colitis appears.

assessments the joint distribution can be represented in a suitable decomposed form and, given any evidence on any node in the graph, conditional probabilities of other nodes can be obtained by the procedures outlined in §6.

(b) *Quantitative conditional probability distributions*

This brings us to the second stage of specifying quantitative conditional probability distributions for each variable v given each configuration of its parent nodes. In genetics these may be generated by known inheritance laws, but in general they may be considered parameters of the system that need to be either estimated from data or subjectively assessed. It is clearly essential to use relevant data if it is available, but in this context no analytic study has apparently been performed. We therefore consider specific models for each of the terms in (4), and provide some numerical assessments based on the judgement of Dr Burdon.

In our application the assessments have been specified to point values as if they were precisely known, but Spiegelhalter & Lauritzen (1990) show how the conditional probabilities can themselves be considered as random quantities that form an additional layer on the graph, and formal bayesian analysis then allows updating of the distributions on those probabilities as data accumulates over a set of cases. In general we would view subjective judgements as essentially providing a 'head-start' in order to begin a system, and Spiegelhalter & Cowell (1991) show that, if the subjective assessments are assumed to be fairly imprecise then they are rapidly overwhelmed by accumulating data. The possibility of adopting such a 'learning' procedure within adverse drug reactions is touched upon in our conclusions.

Node X: day of first relevant exposure to C. difficile. We assume that the first possible relevant day of exposure is $T+1$ and, in the absence of evidence to the contrary, assume that thereafter daily exposures are independent events with probability $\lambda(d)$

on day d, which may depend on whether the patient is in hospital (H). Hence the distribution of the day of first exposure is assumed to be a geometric distribution

$$p(X = d \mid H) = \lambda(d) \prod_{t=T+1}^{d-1} \{1-\lambda(t)\} \quad (d = T+1, T+2, \ldots).$$

Initial values for the daily risks have been set at

$$\lambda(d) = \begin{cases} 0.0002, & \text{if patient is outside hospital on day } d, \\ 0.01, & \text{if patient is in hospital on day } d, \\ 0.9, & \text{if there is a known epidemic of } C.\ difficile \text{ on day } d \text{ in the hospital.} \end{cases}$$

Node S: day of susceptibility of overgrowth. We assume that susceptibility to overgrowth occurs with probability σ, and if it does occur then is on day 1. Hence the distribution for S is

$$p(S = T+1 \mid T) = \sigma,$$
$$p(S = \infty \mid T) = 1-\sigma.$$

A typical value for σ might be 0.9. This quantity is the essential determinant of the adverse risk associated with the drug, and we shall later discuss the issue of learning about this crucial parameter from epidemiological data.

Node E: the day susceptibility ends. If susceptibility has occurred, then it is thought to be certain to continue for a period, say g days, after the drug has been stopped (day W). After this, the susceptibility may or may not continue for some time. We have modelled this by a geometric distribution that results from assuming there is a chance m, known as the susceptibility stop-rate, of ceasing to be susceptible on any day. Thus the distribution for E is given by

$$\text{if susceptible;} \quad p(E = d \mid S = T+1, W) = m(1-m)^{d-W-g-1}$$
$$(d = W+g+1, W+g+2, \ldots);$$
$$\text{if not susceptible;} \quad p(E = \infty \mid S = \infty, W) = 1.$$

After extensive discussion, values of $g = 5$ and $m = 0.05$ have been assessed, which gives an expectation of about 3.5 weeks for susceptibility continuing after the drug is stopped.

Node O: the day overgrowth occurs. This is assumed to follow logically as soon as relevant exposure occurs while the patient is susceptible, and to happen on the day on which this combination of circumstances first occurs. Thus O is defined logically in terms of its influences X, S and E, and to have distribution as follows:

$$\text{if susceptible;} \quad p(O = d \mid X = d, S = T+1, E) = \begin{cases} 1 & \text{if } T+1 \leqslant d \leqslant E, \\ 0 & \text{if } E < d. \end{cases}$$
$$\text{if not susceptible;} \quad p(O = \infty \mid X, S = \infty, E = \infty) = 1.$$

Node C: day colitis appears. There is a lag of a few days between overgrowth occurring and the symptoms of colitis becoming apparent, and we assume a discrete distribution

$$p(C = d \mid O) = p(d-O) \quad (d = O+1, O+2, \ldots).$$

A plausible distribution assumes $p(2) = 0.47$, $p(3) = 0.47$, $p(4) = 0.05$, $p(5) = 0.01$.

5. The two-drug model

We now consider the graphical representation of the two-drug model shown in figure 4 (the background parameters have been left implicit). This has two highlighted components that exactly match the one-drug model considered above, in that an exposure process interacts with a susceptibility process to lead to overgrowth. However, the individual models for the two drugs do meet in two aspects: they have a common exposure process (shown at the top of the graph) and they have a possible common consequence in the development of colitis. We now step through the graph and provide details of the dependencies and any additional numerical assessments. It should be clear that the representation shown below was not immediately apparent, and modelling the process with an appropriate graphical structure took some time and many false starts.

Nodes Y, X_1 and X_2: the exposure process. The graphical formulation of the exposure process expresses that relevant exposure may occur for the first drug before the second is started, but if it does not then any subsequent exposure is relevant to both drugs simultaneously. Hence Y and X_2 are independent variables that summarize first relevant exposure within two disjoint intervals, and X_1 is logically derived in that $X_1 = Y$ if Y is not infinity, and $X_1 = X_2$ otherwise. Formally, we have that

$$p(Y=d\,|\,H) = \lambda(d) \prod_{t=T_1+1}^{d-1} \{1-\lambda(t)\} \quad (d = T_1+1, T_1+2, \ldots, T_2),$$

$$p(Y=\infty\,|\,H) = \prod_{t=T_1+1}^{T_2} \{1-\lambda(t)\},$$

$$p(X_2=d\,|\,H) = \lambda(d) \prod_{t=T_2+1}^{d-1} \{1-\lambda(t)\} \quad (d = T_2+1, T_2+2, \ldots).$$

If $\qquad Y \neq \infty, \quad p(X_1 = d\,|\,Y=d, X_2) = 1.$

If $\qquad Y = \infty, \quad p(X_1 = d\,|\,Y=\infty, X_2 = d) = 1,$

which simply expresses that $X_1 = \min\{Y, X_2\}$. The numerical values for the exposure process are as in the one-drug model.

Nodes S_1, E_1, O_1, S_2, E_2, O_2: the individual drug models. These are as described for the one-drug model adapted to the particular times for taking each drug, but possibly using drug-specific assessments for the susceptibility σ and the delay m in losing susceptibility.

Nodes O, I_1 and I_2: the overgrowth process. Nodes O_1 and O_2 represent the potential for causing overgrowth; to model the actual overgrowth that leads to symptoms we need to define O as the minimum of O_1 and O_2. The indicator variables I_1 and I_2 then show which of the drugs was responsible for the overgrowth, allowing for the fact that it is feasible that both windows of susceptibility were open at the common time of relevant exposure $X_1 = X_2$; hence $O_1 = O_2$ and both drugs could be held equally responsible. Formally, we have that

$$O = \min\{O_1, O_2\}; \qquad I_1 = \begin{cases} 1 & \text{if } O = O_1, \\ 0 & \text{otherwise}; \end{cases} \qquad I_2 = \begin{cases} 1 & \text{if } O = O_2, \\ 0 & \text{otherwise}. \end{cases}$$

The conditional probability tables for O, I_1 and I_2 are then degenerate with ones in the appropriate positions.

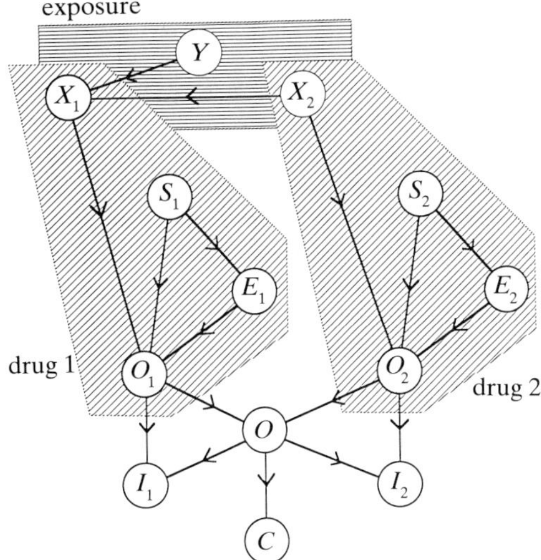

Figure 4. Graphical model of process when two drugs A and B have been administered, starting on days T_1 and T_2 respectively. Exposure: Y, day of first exposure unique to first drug (i.e. between T_1+1 and T_2 inclusive); X_1, day of first exposure relevant to first drug (i.e. after T_1+1); X_2, day of first exposure relevant to second drug (i.e. after T_2+1). First drug: S_1, day of start of susceptibility due to first drug; E_1, day susceptibility due to first drug ends; O_1, day of potential overgrowth due to first drug. Second drug: S_2, day of start of susceptibility due to second drug; E_2, day susceptibility due to second drug ends; O_2, day of potential overgrowth due to second drug. Causation: O, day of actual overgrowth; I_1, actual overgrowth caused by first drug; I_2, actual overgrowth caused by second drug. Case findings: C, day colitis appears.

Node C: the colitis symptoms. This is exactly as for the one-drug model.

We note that this representation allows a number of different interpretations for 'causation' to be simultaneously considered. If we observe colitis to have occurred on a particular day C, and then interrogate, say, I_1, we find the probability of causation by drug A in the sense that exactly the same event would have occurred had drug B not been given. Since $p(I_2 \mid C)$ gives the probability that the exact adverse event would have occurred if B alone had been given, we have that $p(I_1 \mid C)+p(I_2 \mid C)-1$ is the probability of 'joint causation', in the sense that at first relevant exposure both drugs would have led to the window of susceptibility being open. Thus we can decompose the blame into three components:

$1-p(I_2 \mid C)$: the probability that A alone caused the reaction,

$1-p(I_1 \mid C)$: the probability that B alone caused the reaction,

$p(I_1 \mid C)+p(I_2 \mid C)-1$: the probability of joint causation.

Alternatively, the probability that O_1 is non-zero represents the chance that drug A would have led to colitis at some point in the future, even if the actual overgrowth observed was due to drug B. Finally, we could examine the probability of S_1, which represents the event that drug A made the patient susceptible to overgrowth, even if in this case the patient was fortunately not exposed in the appropriate interval. Each of these events could be argued to be relevant to 'causation', and an advantage of a realistic model is that a unique definition need not be sought.

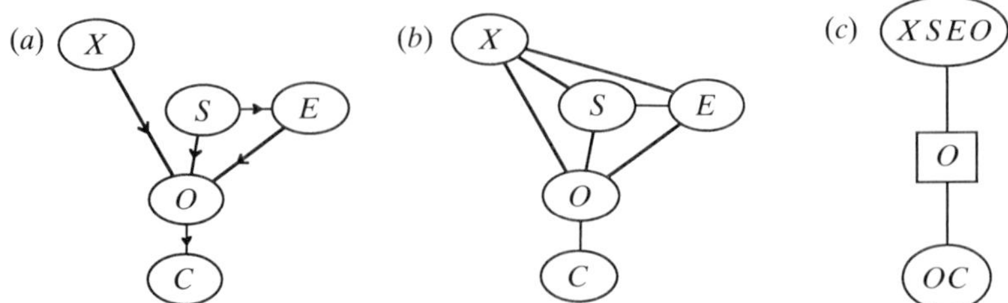

Figure 5. States of graphical re-representation for the one-drug problem. The original network (*a*) is turned into a moral graph (*b*) by adding links between co-parents, and dropping directions. Since this graph is already triangulated, its cliques can be arranged as a clique-tree (*c*).

6. Computational techniques based on graphical models

Having formally specified the joint distribution as the product of terms (4), we need to be able to calculate efficiently quantities of interest, such as the marginal probability of colitis occurring, or the chance of causation by drug A given colitis is observed on a particular day. The representation (4) is not appropriate for such calculations, since it would not be feasible to perform the multiplication and store the full joint distribution in the computer memory. Pearl (1986) showed how, if the graph had the structure of a tree (no cycle in the undirected graph formed by ignoring the directions on the links), then the required results could be obtained using only 'local' computations without specifying the full distribution. Furthermore, there is no need to have an overall controller of the process; each node communicates autonomously with its neighbours in the graph. Problems arise when loops occur in the graph, as in our examples, and Lauritzen & Spiegelhalter (1988) showed that local computations can still be used provided a new representation is found for the joint distribution that is related to an undirected graph, derived from the original DAG, but with a specific property. This stage of graphical re-representation requires five steps:

1. Add an undirected link between all co-parents that are currently unjoined.
2. Drop all directions in the graph. (This forms the so-called 'moral' graph.)
3. 'Triangulate' the graph by filling-in sufficient additional links to ensure that there is no cycle of length 4 or more without a short-cut.
4. Identify the cliques of this triangulated graph as the maximal sets of nodes that are all neighbours of each other.
5. Join the cliques together as a tree, which has the special property that if a node v is contained in any two cliques C and D, then v is also contained in all cliques on the unique path between C and D in the tree.

Those unfamiliar with this area may be somewhat mystified by such a sequence of graph-theoretic operations, and we try to give a short probabilistic justification as we step through the examples: see, for example, Lauritzen & Spiegelhalter (1988), Jensen *et al.* (1990), Andersen *et al.* (1989), Shenoy & Shafer (1988) and Dawid (1991) for further theoretical exposition.

Figures 5 and 6 show these five steps of graphical re-representation carried out to the one-drug and two-drug model. After steps (1) and (2) we are left with an undirected graph whose cliques (sets of nodes who are all joint neighbours) comprise a node and its parents. Referring to expression (3) we see that the joint density is a product of terms defined on this clique set. We therefore have a Markov random field defined on the moral graph, and we may use the independence properties derived

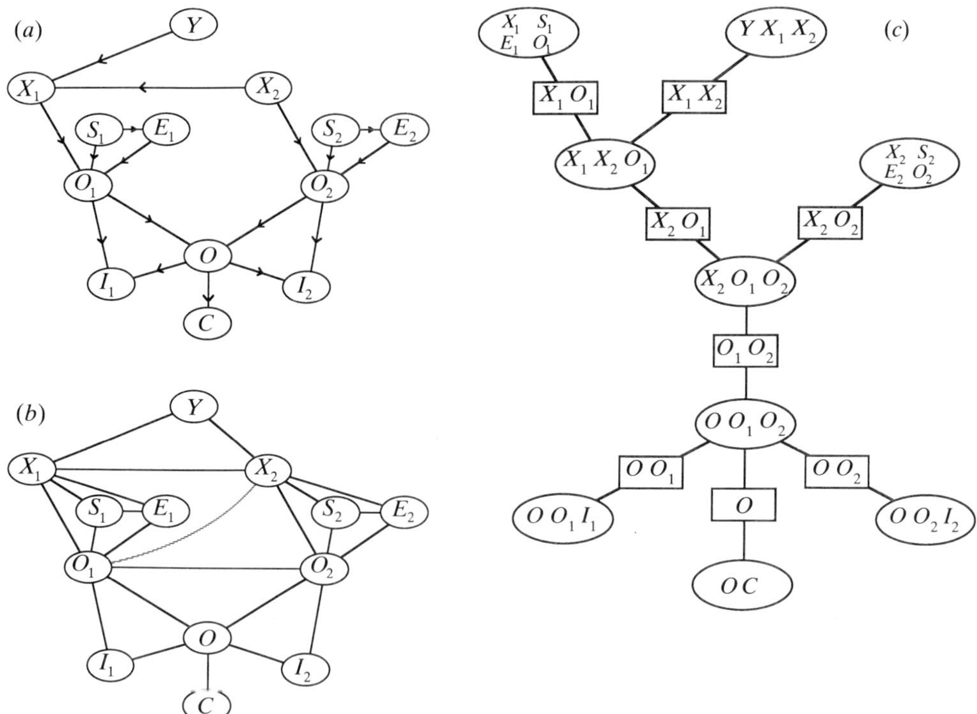

Figure 6. Graphical re-representation of the two-drug model (*a*). In this case an additional link is required to triangulate the moral graph (*b*), which then allows the cliques to be arranged as a clique-tree (*c*).

from undirected graphs that are more usually exploited in applications such as image processing (Isham 1981). More relevantly, there is a growing interest in graphical representations for contingency table models (Edwards & Kreiner 1983). Darroch *et al.* (1980) show how decomposable models (those whose joint distributions can be expressed wholly in terms of the marginal distributions on the cliques) correspond to triangulated graphs in the sense described above, and step (3) (filling in additional links to make triangulated) ensures that from now on it will be sufficient to derive the appropriate clique marginals; essentially we have embedded our original structure in one that is more complex, in that some of the conditional independence assumptions are no longer apparent from the graph, but allows much simpler analysis.

The moral graph of the one-drug model requires no triangulation, whereas the two-drug moral graph has a cycle $\{X_1, X_2, O_2, O_1\}$. To short-circuit this cycle a single link has to be introduced, which could be either X_1–O_2 or X_2–O_1 as chosen.

Triangulated graphs have been extensively explored in the relational database literature (Tarjan & Yannakakis 1984) and it is known that the cliques (step (4)) of such a graph have the powerful property that a join-tree, as defined in step (5), always exists. This means that when evidence arrives concerning any particular node, its implication can be propagated to any other node through the tree without any global supervision to prevent inconsistencies; each clique can act autonomously in receiving and passing on 'messages' from its neighbours. The clique tree for the two-drug model illustrates this property well: node O_1 occurs in five cliques, but these form a chain in the tree. Each clique in the triangulated graph forms a node in the clique tree, and the separator sets are the clique intersections through which the messages pass. We note that each of the separator sets also separates the original directed graph into two or more components.

Before considering the exact form of the messages, we need to complete the next stage in getting the network ready to receive evidence; whereas the preceding stage involved the qualitative graph, we now need to consider the quantitative aspects. We have seen how the joint distribution was originally expressed as a product of functions on nodes and their parents, and we now need to express it as a product of functions ψ and ϕ defined on the cliques and their separators respectively. These functions are called potentials and we shall see below how this representation facilitates the required computations. Specifically, we require a joint distribution expressed as

$$p(V) = \prod_{C \in \boldsymbol{C}} \psi(v_C) \Big/ \prod_{S \in \boldsymbol{S}} \phi(v_S), \tag{5}$$

where $\boldsymbol{C}$ and $\boldsymbol{S}$ are the sets of cliques and separators respectively, and v_C and v_S indicate nodes in sets C and S respectively. Now, since the model is decomposable we have that a particular potential representation takes the form

$$p(V) = \prod_{C \in \boldsymbol{C}} p(v_C) \Big/ \prod p(v_S), \tag{6}$$

where p indicates the marginal distributions on the cliques and separators. From this representation the marginal distributions on any single node v may be easily obtained from the distribution on any clique which contains v. The essence of our approach is a general algorithm for going from any potential representation (5) to a clique marginal representation (6); once this is available we need only show that the representation (5) is not only easily obtained from the original conditional probability form (3), but also that the form (5) is retained when evidence on any nodes is elicited. We shall first show that the potential representation (5) is present at initialization and is retained under conditioning, and then describe the propagation algorithm for deriving (6) from (5).

Comparison of expressions (3) and (5) shows that the potential representation is straightforward to achieve at initialization: we need only set the original clique potentials (v_C) as the product of conditional probability tables corresponding to any node which, with its parents, lies in the clique. For example, in the one drug model the clique potential on $\{X, S, E, O\}$ is initialized to be $p(X)\,p(S)\,p(E \mid S)\,p(O \mid X, S, E)$, while in the two drug model the potential on the clique $\{O_1, O_2, O\}$ is set to $p(O \mid O_1, O_2)$. For many cliques, for example $\{X_2, O_1, O_2\}$, the initializing potential function can be considered to be unity since no node and its parents lie in that clique. All potentials $\phi(v_S)$ on clique separators are set to unity.

We now have a joint distribution expressed in terms of functions on cliques and their separators, and the next step is to show that this representation will hold whatever evidence is received on the network, i.e. for any observation $W = w$, $p(V \backslash W \mid W = w)$ can be expressed in the form (5). This is not difficult to see, since we always have $p(V \backslash W \mid W = w) \propto p(V \backslash W, W = w)$; thus to absorb evidence a clique containing the node that has been observed is selected and any potential not defined on the observed value is set to zero. This may occur at various parts of the network before propagation takes place.

We now consider the propagation algorithm. First, a root clique is arbitrarily

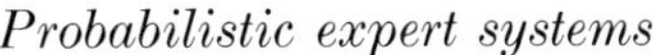

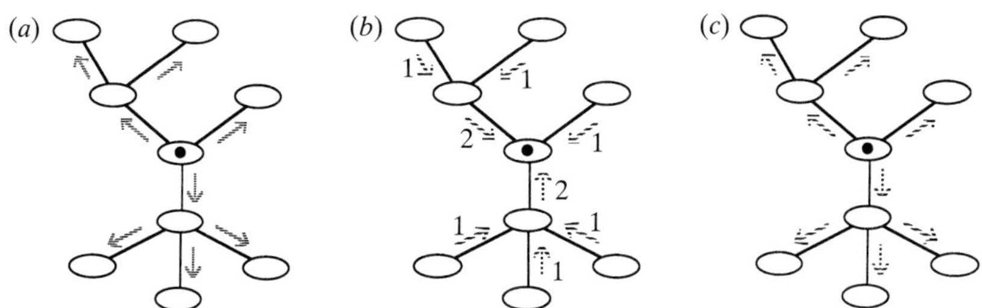

Figure 7. Sequence of message-passing in clique tree for two-drug model. The root-clique sends out a message to 'collect evidence', and then evidence is propagated back to the root. Finally, the collected evidence is distributed through the tree, giving the correct clique posterior distributions which can be marginalized to their constituent nodes.

selected in the clique-tree. It sends out a message 'collect evidence' to its neighbours, who communicate it to their neighbours until it reaches the ends of the tree. Each clique then collects evidence from those neighbours further from the root clique, using the fundamental operations shown below for passing a message from clique C to clique D through separator S.

(i) Marginalize $\psi(v_C)$ to $\phi'(v_S) = \Sigma_{v_{C\setminus S}} \psi(v_C)$.

(ii) Multiply $\psi(v_D)$ pointwise by $\phi'(v_S)/\phi(v_S)$.

(iii) Replace $\phi(v_S)$ by $\phi'(v_S)$.

The process of message-passing is shown in figure 7, which emphasizes that in collecting evidence each clique must await the messages from its more distant neighbours before passing on towards the root. When finally the root clique has collected evidence from its immediate neighbours, it normalizes its new potential to add to unity; this is the correct marginal distribution for the root clique. It then distributes evidence, which involves the same message-passing operation but working back through the tree. When this is complete all cliques and separators will hold their correct marginal distributions from which the current distribution for any node can be easily derived. The algorithm described above is a simplification of that described in Lauritzen & Spiegelhalter (1988), and proofs that these operations lead to this conclusion may be found in Jensen *et al.* (1990); Shenoy & Shafer (1988) provide an axiomatic basis for a general propation scheme, while Dawid (1991) provides numerical examples and generalizes the algorithm to solve many related problems on networks.

The above procedure allows for many facilities, such as easily retracting the items of evidence, exploring influential observations, and planning future questions (Lauritzen & Spiegelhalter 1988). Moreover, some simplifications are possible within specific circumstances. For example, once an initial clique marginal representation has been obtained, the effect of single items of evidence may be propagated using only distribute evidence and hence only a single pass through the network. Also, if only certain nodes of the network are of interest, it may be only necessary to propagate single items of evidence the necessary distance; for example, in the two-drug network, from a clique containing C to the first cliques containing I_1 and I_2. Finally, we could perform sensitivity analyses by varying the quantitative assessments along plausible ranges in order to calculate intervals for the probabilities of interest.

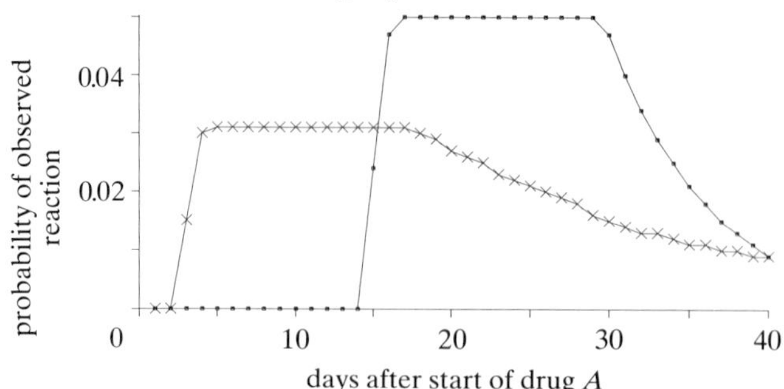

Figure 8. For each drug given considered singly, the probability of colitis occurring on each day, given it occurs at some time. ×, Drug A alone; □, drug B alone.

7. Examples

Our first example was introduced in §3. For illustrative purposes, we assume the two drugs have some distinct characteristics: while drug A is as described in §4, with susceptibility risk $\sigma_A = 0.9$ and susceptibility stop-rate $m_A = 0.05$ per day, drug B has susceptibility risk $\sigma_B = 0.2$ and susceptibility stop-rate $m_B = 0.15$ per day. Otherwise the parameters are taken to be those given in §4. Each drug is first considered separately with the one-drug model, from which we calculate that their prospective probabilities of causing PMC are 0.0058 and 0.0008, respectively, whose ratio 7.25 provides the prior odds on A against B causing a reaction (expression (1) in §2). Conditionally on causing a reaction at some currently unspecified time, the predictive distributions of C, the day of colitis, are shown in figure 8; the distributions are very flat until about a week after the drug is stopped, when they begin to tail off. Clearly, changing relevant parameters could shorten or extend this period. These conditional distributions have ordinates 0.019 and 0.050 for day 27, giving a likelihood ratio of 0.38 for Bayes's theorem (expression (2) in §2). Therefore, if we were to assume that one and only one of the drugs was responsible, our posterior odds on A against B given the case-specific evidence would be $0.38 \times 7.25 = 2.76$, or a probability of 0.73 for A. These posterior odds could also have been obtained directly as the ratio of the unconditional predictive probabilities of colitis occurring exactly on day 27 under each drug given singly.

However, a more realistic analysis exploits the two-drug model to allow for the possibility of joint responsibility. When we condition on knowing $C = 27$, this evidence propagates through the network using the techniques described in the previous section to provide posterior probabilities $p(I_1 \mid C = 27) = 0.862$ and $p(I_2 \mid C = 27) = 0.310$, which decompose into probabilities of 0.690 on A alone, 0.138 on B alone, and 0.172 on joint causation.

These conclusions may, however, be sensitive to information available on the hospitalization of the patient. Suppose, say, that it is known that the patient spent from days 5 to 10 in hospital, although no epidemic of *C. difficile* has been reported. Including this information in the two-drug model leads to the probabilities of causation shown in table 2. This very slightly shifts the blame to drug B. However, this is greatly magnified if we suppose that an epidemic has occurred in the hospital. This change is explained by examining the distribution on other nodes in the network that reveal, for example, that in these circumstances $p(S_1 \mid C = 27) = 0.0003$; i.e. that the first drug almost certainly did not induce susceptibility, even though it was of high risk of doing so. This becomes a more intuitive conclusion when one realizes

Table 2. *Probabilities of causation under different assumptions*

	drug A alone	drug B alone	both A and B
assuming no joint causation (two one-drug models)	0.73	0.27	—
two-drug model	0.690	0.138	0.172
hospitalized days 5–10	0.689	0.139	0.172
exposed to epidemic days 5–10	0.0003	0.9997	0

that, if drug A had induced susceptibility, then in such a high-risk environment overgrowth and colitis would almost certainly have occurred very much earlier than was actually observed.

8. Discussion and conclusions

The models described above have been programmed in C on a personal computer (Cowell *et al.* 1991), and the program allows interactive examination of the effect of changing the context and assumptions concerning the relevant parameters that influence the conditional probability tables. The program is currently being evaluated by the drug surveillance unit of a pharmaceutical company and, although no formal evaluation has taken place, examples such as that above have been presented to both expert assessors of ADRs and to microbiologists, who felt that it mirrored their considered judgement. The aim of such systems is to formalize and make available the expertise of the microbiologist in the routine assessment of cases, and so a formal peer review study is one form of evaluation. A more statistical criterion is whether or not the predictions made by the model accurately reflect the risks of developing the disease, and it is clear that some additional factors will need to be taken into account to make the model adequately realistic.

One issue is blocking, as mentioned in §2, in which a drug may induce susceptibility but might still prevent overgrowth while it is being taken. The tendency for this to occur depends on both the specific drug and the route of administration, with a higher chance of blocking if given intravenously rather than orally. This phenomenon has the effect of making irrelevant any exposure to *C. difficile* while the drug is being taken, and so is easily handled within our framework by introducing a 'blocking' node as a direct influence on the exposure node X. The chance of the blocking node being positive is influenced by the drug and route of administration, and if negative has no effect on X, while if positive shifts to zero the exposure rate (d) for d between day T of starting the drug and day W of withdrawal. The effect of introducing this node is to adjust our probabilities of causation for the possibility that one or other of the drugs was blocking overgrowth.

Adequacy of the model's predictions depends both on the qualitative structure, about which we feel fairly confident, and the assessed parameters, which may be open to more argument. This naturally brings us to the crucial issue of learning about the parameters as data on a number of cases accumulate. A parameter of vital interest to pharmaceutical companies is the susceptibility rate σ, whose size is the main determinant of the safety of the drug with respect to this particular reaction. If we acknowledge the uncertainty concerning this parameter, then it is possible to consider σ a random quantity and include it in the graph as a parent node of S, the day of susceptibility. Spiegelhalter & Lauritzen (1990) describe how one can then process a case using the current expected value of σ, which is essentially the current 'best guess', but then revise our belief about σ after case evidence has been obtained. This revised belief is then carried over to the next case when a new estimate of σ is used for processing.

While this sequential updating process is attractive and is immediately appropriate in contexts such as clinical diagnosis, there are a number of problems with using it in the context of adverse reactions. The first problem is that we may not have independent confirmation of the cause of the reaction, and cannot be definitive that the drug caused susceptibility. While the formal bayesian updating procedure should still update the susceptibility parameter to the appropriate extent, it might be rather unsatisfactory to rest the condemnation of a drug on a set of cases in which it had not definitely been found to be at fault. The second, and perhaps more important problem, is the fact that a drug surveillance unit only hears of a highly selected sample of cases, and that the reporting process could effect our opinion concerning a particular drug. We are currently exploring the possibility of modelling the reporting procedure, with the aim of examining the sensitivity of our conclusions to reporting biases. One way in which a drug might be naively condemned due to circumstantial evidence is when two drugs are routinely given in the same order, and the second drug then will generally appear to have been more likely to have been the causal agent, even if they had identical susceptibilities. We are confident that a formal learning procedure will not be misled into systematically increasing the susceptibility parameter of the second drug.

The models described in this paper have been developed incrementally, and increasing realism has been accompanied by inevitable additional complexity. However, we have been impressed by the way in which the original network formalism has been able to adapt in a modular fashion to these demands, and we feel confident that problems such as those outlined above can be incorporated into the formalism. In addition, we are now attempting to model other classes of reaction along similar lines. While laying no claim as to the absolute correctness of either the qualitative or quantitative components of the models, we hope that their explicitness points the way to a more coherent analysis of many classes of adverse drug reactions.

We are indebted to Douglas Burdon and Sue Roden for their contribution to the project. Stephen Senn made many valuable criticisms. Financial support was provided by Glaxo Group Research Ltd and the Science and Engineering Research Council.

References

Andersen, S. K., Olesen, K. G., Jensen, F. V. & Jensen, F. 1989 HUGIN – a shell for building Bayesian belief universes for expert systems. In *Proceedings of 11th International Joint Conference on Artificial Intelligence, Detroit*, pp. 1080–1085. San Mateo: Morgan Kaufman.

Andreassen, S., Hovorka, R., Benn, J., Olesen, K. G. & Carson, E. 1991 A model-based approach to insulin adjustment. In *Proc. 3rd Eur. Conf. on Artificial Intelligence in Medicine, Maastricht.*

Andreassen, S., Jensen, F. V., Andersen, S. K., Falck, B., Kjaerulff, U., Woldbye, M., Sorensen, A. R., Rosenfalck, A. & Jensen, F. 1989 MUNIN – an expert EMG assistant. In *Computer-aided electromyography and expert systems* (ed. J. E. Desmedt), pp. 255–277. Elsevier.

Beinlich, I. A., Suermondt, H. J., Chavez, R. M. & Cooper, G. F. 1989 The ALARM monitoring system: a case study with two probabilistic inferencing techniques for belief networks. In *AIME '89: Lecture Notes in Medical Informatics vol. 38* (ed. J. Hunger, J. Cookson & J. Wyatt), pp. 247–258. Berlin: Springer-Verlag.

Cowell, R. G., Dawid, A. P., Hutchinson, T. A. & Spiegelhalter, D. J. 1991 A Bayesian expert system for the analysis of an adverse drug reaction. *Artif. Intell. Med.* **3**, 257–270.

Darroch, J. N., Lauritzen, S. L. & Speed, T. P. 1980 Markov fields and log-linear models for contingency tables. *Ann. Statist.* **8**, 522–539.

Dawid, A. P. 1991 Applications of a general propagation algorithm for probabilistic expert systems. *Statist. Computing.* (In the press.)

Edwards, D. & Kreiner, S. 1983 The analysis of contingency tables by graphical models. *Biometrika* **70**, 553–562.

Hutchinson, T. A. & Lane, D. A. 1989 Assessing methods for causality assessment of suspected adverse drug reactions. *J. Clin. Epidemiol.* **42**, 5–16.

Hutchinson, T. A., Dawid, A. P., Spiegelhalter, D. J., Cowell, R. G. & Roden, S. 1991*a* Computer aids for probabilistic assessment of drug safety. I. A spread-sheet program. *Drug Inf. J.* **25**, 29–39.

Hutchinson, T. A., Dawid, A. P., Spiegelhalter, D. J., Cowell, R. G. & Roden, S. 1991*b* Computer aids for probabilistic assessment of drug safety. II. An expert system. *Drug Inf. J.* **25**, 41–48.

Isham, V. 1981 An introduction to spatial point processes and Markov random fields. *Int. statist. Rev.* **49**, 21–43.

Jensen, F. V., Lauritzen, S. L. & Olesen, K. G. 1990 Bayesian updating in recursive graphical models by local computations. *Comp. Statist. Q.* **4**, 269–282.

Karch, F., Smith, C., Kerzner, B., Mazullo, J., Weintraub, M. & Lasagna, D. 1976 Adverse drug reactions – a matter of opinion. *Clin. Pharmac. Therap.* **19**, 489–492.

Lane, D. A. 1989 Subjective probability and causality assessment. *Appl. stoch. Models Data Analysis* **5**, 53–76.

Lane, D. A., Kramer, M. S., Hutchinson, T. A., Jones, J. K. & Naranjo, C. 1987 The causality assessment of adverse drug reactions using a Bayesian approach. *Pharm. Med.* **2**, 265–283.

Lauritzen, S. L. & Spiegelhalter, D. J. 1988 Local computations with probabilities on graphical structures and their application to expert systems (with discussion). *Jl R. statist. Soc.* B **50**, 157–224.

Lauritzen, S. L., Dawid, A. P., Larsen, B. N. & Leimer, H.-G. 1990 Independence properties of directed Markov fields. *Networks* **20**, 491–505.

Pearl, J. 1986 Fusion, propagation and structuring in belief networks. *Artif. Intell.* **29**, 241–288.

Pearl, J. 1988 *Probabilistic reasoning in intelligent systems.* San Mateo: Morgan Kaufmann.

Proceedings of DIA Workshop 1986 The future of adverse drug reaction diagnosis: computers, clinical judgement and the logic of uncertainty. *Drug Inf. J.* **20**, 455–533.

Shafer, G. & Pearl, J. (eds) 1990 *Readings in uncertain reasoning.* San Mateo: Morgan Kaufmann.

Shenoy, P. P. & Shafer, G. R. 1988 Axioms for probability and belief-function propagation. *Uncertainty in artificial intelligence 4*, pp. 169–198. Amsterdam: North-Holland.

Spiegelhalter, D. J. & Cowell, R. G. 1991 Learning in probabilistic expert systems. In *Bayesian statistics 4*. Valencia University Press.

Spiegelhalter, D. J. & Lauritzen, S. L. 1900 Sequential updating of conditional probabilities on directed graphical structures. *Networks* **20**, 579–605.

Szolovits, P., Patil, R. S. & Schwartz, W. B. 1988 Artificial intelligence in medical diagnosis. *Annals Int. Med.* **108**, 80–87.

Tarjan, R. E. & Yannakakis, M. 1984 Simple linear time algorithms to test chordality of graphs, test acyclicity of hypergraphs, and selectively reduce acyclic hypergraphs. *SIAM J. Computing* **13**, 566–579.

Thompson, E. A. 1986 Genetic epidemiology: a review of the statistical basis. *Statist. Med.* **5**, 291–302.

Turner, W. M. 1984 The Food and Drug Administration algorithm. *Drug Inf. J.* **18**, 259–266.

Venulet, J. 1984 The Ciba-Geigy approach to causality. *Drug Inf. J.* **18**, 315–318.

Wright, S. 1934 The method of path coefficients. *Ann. math. Statist.* **5**, 161–215.

Estimation of parameters in hidden Markov models

By W. Qian and D. M. Titterington
Department of Statistics, University of Glasgow, Glasgow G12 8QQ, U.K.

Parameter estimation from noisy versions of realizations of Markov models is extremely difficult in all but very simple examples. The paper identifies these difficulties, reviews ways of coping with them in practice, and discusses in detail a class of methods with a Monte Carlo flavour. Their performance on simple examples suggests that they should be valuable, practically feasible procedures in the context of a range of otherwise intractable problems. An illustration is provided based on satellite data.

1. Introduction

There is a wide class of problems in statistics that can be regarded as incomplete-data problems. The observed data, to be modelled as a realization of a random vector, Y, are interpreted as a partly observed version of a 'complete' set of data, which would be a realization of a random vector, Z. Often, Z has physical meaning, as in the context of data with missing values. Here

$$Z = (X, Y), \tag{1.1}$$

where X represents the missing values.

It should be emphasized that there is a range of incomplete-data problems that are not so obviously represented by the 'complete data = observed data + missing values' paradigm of (1.1). These include such topics as censored and truncated data; see Dempster *et al.* (1977) and Little & Rubin (1987) for many examples and much discussion. However, in the present paper, we concentrate on missing data problems. In particular, we assume that

$$Z = \{Z_i\},$$

where $Z_i = (X_i, Y_i)$, $i = 1, \ldots, n$. Here, n represents the number of individuals or items in a sample, or, in our later discussion, the number of pixels in an image. Thus each of the n 'observations' consists of multivariate data of which Y_i is observed and X_i is missing. In the present paper, we assume that each X_i corresponds to the same variable or variables in the multivariate data vector. Thus we are considering a more particular structure than is typical in the context of, say, sample-survey data with non-response a possibility in any of the individual variables. In the simplest version of our structure (which is also both very common in practice and the one we shall consider in detail throughout the paper), X_i is one dimensional, discrete and finite valued. Somewhat more generally, each X_i has the same sample space, and similarly for the Y_i. Without too much extra effort, one could cope with additional incompleteness within the Y_i, but we shall not introduce this extra complication.

Phil. Trans. R. Soc. Lond. A (1991) **337**, 407–428
Printed in Great Britain

Some particular examples of our structure are as follows.

(*a*) $\{X_i\}$ are the (unknown) age-categories of n fish and $\{Y_i\}$ are the observed lengths of the fish.

(*b*) $\{X_i\}$ are a time-sequence of n (unknown) configurations of an individual's vocal tract and $\{Y_i\}$ are the corresponding sequence of projected sounds.

(*c*) X denotes the true surface-typing of a pixellated area of land and Y denotes the corresponding image observed by a satellite.

Case (*c*) is just one illustration of a noisy image, other examples of which occur in many fields, such as petrology and medicine; see Aykroyd & Green (1991), for instance.

In all these illustrations, the X_i are 'real' physical quantities. However, in some applications of our models, the X_i are latent variables, familiar in factor analysis and, potentially, in statistical approaches to neural networks; see also §7.

The key feature of our work will be the proposal of a statistical model for Z in the form of the joint probability density

$$f(Z \mid \theta) = f(X, Y \mid \theta), \tag{1.2}$$

where the functional form, f, will be treated as being known. If we are provided with data $Y = y$, therefore, two quantities are uncertain, the associated missing values x and the unknown parameter(s), θ. It pays to recognize the different natures of these two unknown quantities, and this is reflected in the different terminology we shall use for the two operations of trying to identify them on the basis of y; we might wish to (i) impute values for the missing data, x, and/or (ii) estimate the unknown parameter(s), θ.

The importance of making this distinction is emphasized later in the paper and is also highlighted by Little & Rubin (1983). We shall make some reference to objective (i), but the principal aim of this paper is to discuss approaches to (ii) and, in particular, to examine, and attempt to deal with, complications with maximum likelihood estimation in a class of problems with important applications.

The layout of the paper is as follows. In §2, we discuss the class of models of interest and, in §3, we highlight the difficulty of obtaining maximum likelihood estimates of parameters. A discussion of various approaches that attempt to obviate the difficulties in §4 is followed by a detailed, illustrated investigation of a subclass of procedures, based on Monte Carlo methods, in §5. An example involving satellite data is presented in §6 and a small discussion in §7 concludes the paper.

2. Mixture models, dependence structures and applications

There are, formally, the following two ways of factorizing (1.2) in terms of a conditional distribution,

$$f(x, y \mid \theta) = p(y \mid x, \theta)\, \pi(x \mid \theta) \tag{2.1}$$

and

$$f(x, y \mid \theta) = \pi(x \mid y, \theta)\, p(y \mid \theta). \tag{2.2}$$

We generically denote 'densities' associated with X by π, and those associated with Y by p. Clearly, the marginal density for Y, $p(\cdot \mid \theta)$, which defines the likelihood function for the observed data, is also given by

$$p(y \mid \theta) = \int p(y \mid x, \theta)\, \mathrm{d}\Pi(x \mid \theta), \tag{2.3}$$

where $\Pi(\cdot\,|\,\theta)$ is the measure associated with $\pi(\cdot\,|\,\theta)$. Equation (2.3) displays $p(y\,|\,\theta)$ in the form of a mixture density, with $\Pi(\cdot\,|\,\theta)$ as mixing measure and X as the mixing variable. Formulation (2.2) is important, in that the factor $\pi(\cdot\,|\,y, \theta)$ is the natural basis for methods of imputing values for the missing X, given observed data y, but we concentrate largely on (2.1) and (2.3).

It will often be natural to write $\theta = (\phi, \beta)$, where ϕ and β are distinct sets of parameters, each set with its own parameter space, associated respectively with the two factors on the right-hand side of (2.1). Thus

$$f(x, y\,|\,\theta) = p(y\,|\,x, \phi)\pi(x\,|\,\beta). \tag{2.4}$$

Recall that models such as (2.4) are associated with data on n individuals or items. Thus, for instance,

$$(X, Y) = \{(X_i, Y_i), i = 1, \dots, n\}.$$

Very often, the Y_i are conditionally independent, given the X_i, so that

$$p(y\,|\,x, \phi) = \prod_{i=1}^{n} p(y_i\,|\,x, \phi), \tag{2.5}$$

or, even more simply,

$$p(y\,|\,x, \phi) = \prod_{i=1}^{n} p(y_i\,|\,x_i, \phi). \tag{2.6}$$

So far as estimation is concerned, the major complications emanate from the factor $\pi(x\,|\,\beta)$ in (2.4). As described in Titterington (1990), a gradation in complexity can be identified by considering different dependence patterns among the X_i. Except when discussing Example 5.2, we assume throughout that each X_i takes one of a finite set of k qualitative values $\{c_1, \dots, c_k\}$, although some of the formulation can be generalized to cover, for instance, the case of an AR(1) process with additive noise, equivalent to an ARMA(1, 1) process.

(*a*) *Standard mixture model*

Here, the X_i are independent, and we shall assume them to be identically distributed. Thus, if (2.6) obtains, the Z_i are independent and the Y_i are marginally so. As a result,

$$\pi(x\,|\,\beta) = \prod_{i=1}^{n} \pi(x_i\,|\,\beta),$$

and

$$p(y\,|\,\theta) = \prod_{i=1}^{n} \left\{ \sum_{j=1}^{k} p_j(y_i\,|\,\phi)\pi_j \right\}, \tag{2.7}$$

where

$$p_j(y_i\,|\,\phi) = p(y_i\,|\,X_i = c_j, \phi)$$

and

$$\pi_j = \text{prob}\,(X_i = c_j),$$

for all j. (In the context of images, the c_j may be a set of colours, usually representing some discrete classification of the pixels in the true scene.) the π_j are called mixing weights and $p(y\,|\,\theta)$ represents the joint density of a random sample from a finite, k-component mixture distribution. Since the mixing X_i are, marginally, samples from a multinomial distribution, the mixture model may, in this case, be thought of as a hidden multinomial model. There are very many applications of standard mixture models; see Titterington *et al.* (1985), Titterington (1990) and references therein.

(b) Hidden Markov chain model

We now admit a simple form of dependence among the x_i, namely that

$$\pi(x \mid \beta) = \pi(x_1 \mid \beta) \sum_{i=2}^{n} \{\pi(x_i \mid x_{i-1}, \beta)\}.$$

Thus the x_i form a Markov chain, which is generally assumed to be stationary. Since the X_i are now no longer independent, the marginal density for Y_i no longer simplifies in the way that (2.7) does. The Y_i are said to come from a hidden Markov chain model, and the study of such models has become extremely popular in the speech recognition literature (Juang & Rabiner 1991; Bourlard 1990), as well as in the analysis of neurophysiological data.

In the former context, the X_i form a sequence (in time) of underlying prototypical spectra, each representing one of a (finite) number of configurations of the vocal tract, and the Y_i are random functions thereof.

(c) More general hidden Markov models

Perhaps the currently most familiar application of more general hidden Markov models is the hidden Markov random field model for noisy images. Here, the subscript i identifies a pixel-site, or a line-site, in a pixellated image, X represents the true scene and Y the observed image, which is just a blurred and/or noisy version of X.

We allow a more general, but still markovian marginal model for the X_i, and represent their joint density by that of a Gibbs distribution. Thus,

$$\pi(Z \mid \beta) = (1/C(\beta)) \exp\{-U(Z, \beta)\},$$

where $C(\beta)$ is a normalizing constant (the partition function), and the energy function, U, takes the additive form

$$U(Z, \beta) = \sum_{c \in \Theta} V_c(Z, \beta),$$

where Θ is a class of subsets of the sites, and V_c is the potential function associated with subset c. Usually, each $c \in \Theta$ consists of only a few sites. For instance, in the case of pairwise-interaction models, Θ contains only the elementary subsets consisting of the individual sites themselves, along with some two-site subsets, each containing a pair of 'neighbouring' sites; for popular two-dimensional examples, see Geman & Geman (1984). In the context of images, the structure of the Gibbs distribution is intended to reflect plausible local, spatial correlation in the true scene. Thus, typically, each V_c involves only a very small subset of the Z_i. As in the hidden Markov chain case, the likelihood function defined by the marginal density, $p(y \mid \theta)$, does not take a simple form.

The seminal papers on this application are Geman & Geman (1984) and Besag (1986). Those papers and many others, including Aykroyd & Green (1991), contain much material about restoring the scene (equivalent in our phraseology to imputing values for the missing x). Along with Ripley (1988), they also have something to say about the other problem, namely, that of estimating θ, that is the main focus of the present paper.

So far as $p(y \mid x, \phi)$ is concerned, we shall assume that (2.6) holds. Note, however, that (2.6) does not cover the case of systematic blurring, although the general methods we describe are not necessarily limited, in terms of applicability, to non-blurred images.

3. Maximum likelihood estimation: methods and difficulties

As a general rule, parameter estimation from incomplete data is more awkward than would be the case if the data were complete. Often, the additional difficulty is simply that, whereas explicit formulae exist for estimates in the complete-data case, iterative numerical methods are now required. Sometimes, the situation is even more awkward, particularly when the complete-data version is itself non-trivial. The maximum-likelihood treatment of the models in §2 illustrates these points very well; see Titterington (1990) for a more detailed exposition.

We restrict our discussion to the case where each $x_i \in \{c_1, \ldots, c_k\}$.

Even for the simplest (hidden multinominal) model, explicit maximum likelihood estimates are hardly ever (not quite never!) available. Were the x_i known, parameter estimation would often be trivial: the k mixing weights would be estimated by relevant relative frequencies and the parameters associated with, say, $p_j(\cdot \mid x_i, \phi)$ are often easily estimated. Recently, a particular numerical procedure, the EM-algorithm (Dempster *et al.* 1977) has pervaded the incomplete-data literature. In the mth iterative stage, current estimates, $\theta^{(m)}$, are updated to $\theta^{(m+1)}$ by the following double step:

E-step: compute $Q^{(m)}(\theta) \equiv E\{\ln f(X, y \mid \theta) \mid y, \theta^{(m)}\}$,

M-step: find $\theta = \theta^{(m+1)}$ to maximize $Q^{(m)}(\theta)$.

Typically, the M-step is as easy or hard as is the computation of maximum likelihood estimates from complete data; the E-step involves 'averaging out' over the missing values and is sometimes loosely equivalent to imputing for the missing values.

Repetition of the double step generates a non-decreasing sequence of values of the log-likelihood (associated with y) and convergence to a local maximum likelihood estimate can often be proved.

The levels of difficulty involved in the implementation of the EM-algorithm can be summarized as follows, in which 'explicit' means that an explicit formula exists, not requiring numerical integration or summation (E-step) or iterative solution (M-step).

(i) *Hidden multinomial.* E-step: explicit; M-step: explicit.

(ii) *Hidden Markov chain.* E-step: explicit, although it does require a forwards and a backwards recursion through the data; M-step: explicit.

Details are provided in Titterington (1990). Note that, in claiming that the M-step is explicit, it is assumed that we are dealing with a case in which the 'colour-conditional' densities p_j admit explicit maximum likelihood estimates. This is true, for instance, for the case of gaussian densities, but not for beta densities.

(iii) *Hidden Markov random field.* E-step: very difficult; M-step: very difficult.

In the above, 'very difficult' borders on the impossible, at least so far as exact calculation is concerned! To explain this, it is helpful to introduce a slight change of notation. Let X_i now denote an indicator vector, of length k, which has unity as the jth element, and zero elsewhere, if pixel i is of colour c_j. Let X_{ij} denote the jth element of X_i. Then

$$\ln f(X, y \mid \theta) = -\sum_c V_c(X, \beta) - \ln C(\beta) + \sum_i \sum_j X_{ij} \ln p_j(y_i \mid \phi).$$

In the E-step we must therefore compute, given $\theta^{(m)}$,

$$Q^{(m)}(\theta) = -\sum W_c^{(m)}(\beta) - \ln C(\beta) + \sum_i \sum_j X_{ij}^{(m)} \ln p_j(y_i \mid \phi),$$

where
$$W_c^{(m)}(\beta) = E\{V_c(X, \beta) \mid y, \theta^{(m)}\} \tag{3.1}$$
and
$$X_{ij}^{(m)} = E(X_{ij} \mid y, \theta^{(m)}). \tag{3.2}$$

Exact computation of both (3.1) and (3.2) is, typically, impossible.

So far as the M-step is concerned, computation of $\phi^{(m+1)}$ is often straightforward, as in the earlier models. Maximization of the part of $Q^{(m)}(\theta)$ depending on β is not easy, in general, largely because computation of $C(\beta)$ is hardly ever feasible. In other words, even if we have a pure realization from a Markov random field, maximum likelihood estimation of the underlying parameter, β, is not a practical proposition.

The inherent difficulties are well illustrated by the following simple example.

Example 3.1. Exponential family case

Suppose
$$f(x, y \mid \theta) = (1/D(\theta)) \exp\{H(x, y)^{\mathrm{T}} \theta\}. \tag{3.3}$$

Then the EM double step is as follows.

E-step: evaluate $E\{H(X, y) \mid y, \theta^{(m)}\}$.

M-step: solve $E\{H(X, y) \mid \theta\} = E\{H(X, y) \mid y, \theta^{(m)}\}$ to obtain $\theta = \theta^{(m+1)}$.

In general, neither of the above expectations can be computed explicitly, nor can the equation in the M-step be solved without numerical methods. Both expectations can be approximated by sample means created by multiple implementations of the two associated Gibbs samplers. (These Gibbs samplers are simply mechanisms for simulating realizations from the relevant Gibbs distributions. From an arbitrary initial configuration, a realization for a given site is simulated from its distribution conditional on the rest of the configuration. The configuration is updated, the procedure is repeated for each site and the whole operation iterated a 'large' number of times. Ultimately, a valid realization is created (see also Smith 1991).)

However, only in very simple cases is the procedure feasible in practice, because of the scale of computations. Geman & McClure (1987) describe a one-parameter problem, for which an off-line approximation to the function $E\{H(X, y) \mid \theta\}$ is computed at the outset, using a grid of θ-values.

One important idea with considerable promise, also involving Monte Carlo procedures, is the approach of Geyer & Thompson (1992). Although they concentrate on the case of non-noisy data, their idea is useful in the M-step of the EM algorithm for Example 3.1, where we have to maximize, with respect to θ,

$$E\{H(X, y) \mid y, \theta^{(m)}\}^{\mathrm{T}} \theta - \ln D(\theta). \tag{3.4}$$

The difficulty lies in the intractability of $D(\theta)$, but Geyer & Thompson (1992) note that, for any θ',

$$D(\theta) = D(\theta') \int \exp\{H(x, y)^{\mathrm{T}} (\theta - \theta')\} \, \mathrm{d}F(x, y \mid \theta'), \tag{3.5}$$

which is simply proportional to a moment generating function, may be approximated by the empirical counterpart

$$D_N(\theta) = D(\theta') \frac{1}{N} \sum_{r=1}^{N} \exp\{H(x_r, y_r)^{\mathrm{T}} (\theta - \theta')\},$$

where $\{(x_r, y_r), r = 1, \dots, N\}$ are realizations from the distribution with density $f(x, y \mid \theta')$. Since (3.4) involves $\ln D(\theta)$, to be approximated by $\ln D_N(\theta)$, rather then $D(\theta)$ itself, it is not necessary to know $D(\theta')$ when computing the maximal θ. Geyer & Thompson (1992) give much more detail, including guidance about the choice of θ'. Clearly, the choice of $\theta' = \theta^{(m)}$ is a natural one when computing $\theta^{(m+1)}$ in the EM algorithm. Alternatively, one could use the same N simulated realizations throughout the EM algorithm.

As noted by Besag (1976), (3.5) has a history going back at least to Bartlett (1971).

4. Practical parameter estimation for hidden Markov models

A variety of more practicable procedures have been considered for dealing with data from hidden Markov models. Some of them are comparatively general in scope, whereas others are appropriate only for special cases.

4.1. *Methods based on decision-directed imputation*

Recall that, for most hidden Markov models, there are two difficulties: (*a*) X is missing; (*b*) even were X provided, maximum likelihood estimation of β would be difficult.

One general approach is to generate a sequence $\{(\theta^{(m)}, x^{(m)}), m = 0, 1, \dots\}$ of pairs of iterates for θ and the missing X such that $x^{(m+1)}$ is created on the basis of y and $\theta^{(m)}$, and $\theta^{(m+1)}$ is 'estimated' from y and $x^{(m+1)}$.

The general form of the iterative stage in what we might call the restoration–maximization (RM) algorithm is as follows, given $\{\theta^{(m)}, x^{(m)}\}$,

R-step: create $x^{(m+1)}$ from $\pi(x \mid y, \theta^{(m)})$;

M-step: choose $\theta = \theta^{(m+1)}$ to maximize

$$p(y \mid x^{(m+1)}, \phi)\, \pi_p(x^{(m+1)} \mid \beta), \tag{4.1}$$

where $\pi_p(x^{(m+1)} \mid \beta)$ is an amenable alternative to $\pi(x^{(m+1)} \mid \beta)$, such as Besag's (1975) pseudo-likelihood, defined by

$$\prod_{i=1}^{n} \pi(x_i^{(m+1)} \mid x_{\partial i}^{(m+1)}, \beta), \tag{4.2}$$

in which $x_{\partial i}$ denotes the values of x on the neighbouring pixels to pixel i. The crucial simplifying feature of the pseudo-likelihood is that the partition function $C(\beta)$ is absent.

There are various specific versions of the R-step, including the following.

R_{ICM} (Besag 1986): apply Besag's (1986) ICM (iterated conditional modes) algorithm to $\pi(\cdot \mid y, \theta^{(m)})$. This iterative procedure, which typically converges quickly, locates an $x^{(m+1)}$ that may be close to the true mode.

R_{MAP}: compute (or attempt to compute) the true mode of $\pi(\cdot \mid y, \theta^{(m)})$, thereby obtaining the maximum *a posteriori* (MAP) restoration. For very special situations (Greig *et al.* 1989), the MAP restoration may be obtained exactly. Otherwise, techniques such as simulated annealing (Geman & Geman 1984) have been proposed which may or may not attain the mode in practice.

There are, however, unsatisfactory aspects of this approach of trying to find a modal restoration and then behaving, in the M-step, as if the restoration is the true scene. We are effectively attempting to maximize $f(x, y \mid \theta)$ simultaneously with respect to the missing x and the unknown θ. The finite mixture (hidden multinomial) version of this approach is well known to lead to biases in the estimates of θ (see, for example, Marriott 1975; Titterington 1984).

Modal restoration is in the same spirit as so-called decision-directed learning, familiar in the engineering literature on unsupervised learning. To explain this concept, let $\Delta(x)$ denote, in general, a randomized rule whereby restoration x is selected with probability $\Delta(x)$, for all x. Decision-directed rules correspond to Δ being degenerate. Thus modal restoration is representable by

$$\Delta(x) = \begin{cases} 1 & \text{if } x \text{ is the posterior mode,} \\ 0 & \text{otherwise.} \end{cases}$$

For further discussion of decision-directed learning, see, for instance, ch. 6 of Titterington *et al.* (1985).

The problem can be attacked by modifying either or both of the R-step and the M-step.

Qian & Titterington (1991) modified the latter as follows. If (2.6) applies, then (4.1), combined with (4.2), gives

$$\prod_{i=1}^{n} \{p(y_i \mid x_i^{(m+1)}, \phi)\, \pi(x_i^{(m+1)} \mid x_{\partial i}^{(m+1)}, \beta)\}.$$

Instead, Qian & Titterington (1991) used

$$\prod_{i=1}^{n} \{p(y_i \mid x_{\partial i}^{(m+1)}, \theta)\}. \tag{4.3}$$

There are various points to make about (4.3). Note that it can be written

$$\prod_{i=1}^{n} \left\{\sum_{x_i} p(y_i \mid x_i, \phi)\, \pi(x_i \mid x_{\partial i}^{(m+1)}, \beta)\right\} = \prod_{i=1}^{n} \left\{\sum_{j=1}^{k} \pi_j^{(m+1,\, i)}(\beta) p_j(y_i \mid \phi)\right\},$$

where, for each i,

$$\pi_j^{(m+1,\, i)}(\beta) = \pi(x_i = j \mid x_{\partial i}^{(m+1)}, \beta), \quad j = 1, \ldots, k. \tag{4.4}$$

In spite of the tortuous notation, it can be seen that the k quantities in (4.4) are a set of mixing weights, and that (4.3) takes the form of a likelihood from independent observations from finite mixtures of the same component densities. The sets of mixing weights vary, but are clearly linked through the common β, and are locally associated, because of the dependence on $x_{\partial i}$. Qian & Titterington (1991) used a few steps of the EM algorithm to maximize (4.3) within the M-step of the RM algorithm. Clearly, within the ith factor of (4.3), the method still treats $x_{\partial i}^{(m+1)}$ as if it were the truth. However, acknowledgement that x_i itself is unknown appears to lead to improved estimates of the parameters and subsequently to robust restoration procedures; see the authors' reply to the discussion of Besag *et al.* (1991).

Qian & Titterington (1991) call their estimation algorithm, based on (4.3), the point-pseudo-likelihood (PPL)-EM algorithm. They also generalized the method.

4.2. *Methods based on probabilistic-teacher imputation*

It is also possible to modify the R-step in ways that are likely to improve the properties of the resulting estimators. Just as the use of restorations such as posterior modes is comparable with decision-directed imputation, so can one parallel the so-called probabilistic-teacher procedures familiar in the unsupervised-learning literature; see ch. 6 of Titterington *et al.* (1985). One essentially chooses, for $x^{(m+1)}$, a realization from $\pi(x \mid y, \theta^{(m)})$. Thus the randomized rule $\Delta(\cdot)$ introduced in §4.1 is non-degenerate and, in this case, is simply $\pi(\cdot \mid y, \theta^{(m)})$. In general, it is much safer to treat such an $x^{(m+1)}$ as if it were the truth than it is in the decision-directed case. However, any convergence properties of $\theta^{(m+1)}$ will be in law rather than, say, in probability. For the latter, further modification is clearly necessary to create a sequence of parameter estimates with suitably ergodic behaviour. Simulation of $x^{(m+1)}$ can be achieved in various ways, by using the Gibbs sampler of Geman & Geman (1984), for example.

One such modified approach is the following. Choose a positive integer T.

R-step: simulate independent samples $\{x_t^{(m+1)}; t = 1, \ldots, T\}$ from $\pi(x \mid y, \theta^{(m+1)})$.

M-step: for each t, find θ_t, based on $x_t^{(m+1)}$, using (4.1) or any desired variant thereof, and take

$$\theta^{(m+1)} = \frac{1}{T} \sum_{t=1}^{T} \theta_t. \tag{4.5}$$

Thus each simulated x_t creates an estimate of θ, and those estimates are then averaged. It will be helpful to consider this algorithm in more detail in the context of a particular example.

Example 4.1

Suppose that there is an explicit method for estimating θ in the M-step of the RM-algorithm, so that

$$\hat{\theta} = g(H(x, y)),$$

for some functions g and H. This would obtain if $f(x, y \mid \theta)$ were of exponential family form (cf. Example 3.1), with invertible likelihood equation given by

$$E\{H(X, Y) \mid \theta\} = H(x, y),$$

where $H(X, Y)$ are the sufficient statistics. Then (4.5) takes the form

$$\theta^{(m+1)} = \frac{1}{T} \sum_{t=1}^{T} g\{H(x_t^{(m+1)}, y)\}.$$

The following represent special cases.

Example 4.1.1

If $T = \infty$, and function g is linear, this takes us back to the EM algorithm, which can therefore be approximated by using the Gibbs sampler with a large but finite value for T. Chalmond (1989) describes a version of this for maximizing the pseudo-likelihood function, rather than the intractable likelihood. A somewhat similar approach was developed by Veijanen (1990).

Example 4.1.2

The case $T = 1$ corresponds to what Celeux & Diebolt (1985) call the Stochastic EM algorithm (SEM) algorithm. In that paper, they apply their method only to the case of standard finite-mixture data, but it is clearly applicable more generally.

In §5, we shall consider in much more detail the implementation and behaviour of these procedures.

Younes (1989) constructs a stochastic gradient algorithm for the case where the joint distribution of X and Y belongs to the exponential family with sufficient statistics $H(X, Y)$. Then the likelihood equation takes the form

$$E\{H(X, Y) \mid \theta\} = E\{H(X, Y) \mid Y = y, \theta\}.$$

Younes's approach is to generate a sequence of iterations $(\theta^{(m)}\}$ according to

$$\theta^{(m+1)} = \theta^{(m)} + [1/(m+1)U]\,[H(X^{(m+1)}, Y^{(m+1)}) - H(X_y^{(m+1)}, y)], \tag{4.6}$$

for $m = 0, 1, \ldots$. In (4.6), U is a positive constant, $(X^{(m+1)}, Y^{(m+1)})$ are sampled from $f(X, Y \mid \theta^{(m)})$, and $X_y^{(m+1)}$ is sampled from $\pi(\cdot \mid y, \theta^{(m)})$. Of course, in principle, such samples themselves involve, if the Gibbs sampler is used, many passes over the frame.

Instead, one can run a version of the algorithm with $X^{(m+1)}$ created by applying a single cycle of the Gibbs sampler, with $\theta = \theta^{(m)}$, from $X^{(m)}$, and similarly for $X_y^{(m+1)}$.

The algorithm is clearly a Monte Carlo variant of the technique known as stochastic approximation. The procedure described in Younes (1989) for hidden Markov random fields is a direct development of his procedure for the noise-free Markov random field (Younes 1988). Convergence properties are somewhat complicated but are described in the original papers. For further work along these lines see Moyeed & Baddeley (1991).

4.3. *Other methods*

In view of the perceived difficulty of maximum likelihood estimation, various other approaches have been explored. The method of moments has been implemented on simple models, by Frigessi & Piccioni (1988, 1990) for two-dimensional Ising models with binary channel noise (two parameters altogether) and by Geman & McClure (1987) for one parameter in the context of single-photon-emission computed tomography (SPECT). Pickard (1987) develops asymptotic maximum likelihood estimation, but this approach is available only for the Ising model. Possolo (1986) and Derin & Elliott (1987) note that, for certain noise-free binary images, conditional logits are linear functions of the parameters. This motivates a least-squares approach to parameter estimation which is further discussed by Gray *et al.* (1991).

We now concentrate on and illustrate the Monte Carlo-based RM algorithm introduced in §4.1.

5. Monte Carlo restoration–estimation algorithms

We begin this section by proposing a more general framework for the RM algorithm described in §4.1. Suppose that, when the complete data, (x, y), are available, we estimate θ by

$$\hat{\theta} = g(x, y). \tag{5.1}$$

The function g may or may not be explicit, so that (5.1) formally includes any method such as maximum likelihood, maximum pseudo-likelihood or moment estimation.

Consider now the following Monte Carlo approach to deal with the case where only $Y = y$ is given. The procedure generates a sequence $\{\theta^{(m)}\}$ of iterates, starting from an initial guess, $\theta^{(0)}$.

R-step (restoration): for specified T, generate samples $x_{(m+1)1}, \ldots, x_{(m+1)T}$ from $\pi(x \mid y, \theta^{(m)})$.

E-step (estimation): two possibilities are considered here,

$$E_A: \qquad \text{take } \theta^{(m+1)} = \frac{1}{T}\sum_{j=1}^{T} g(x_{(m+1)j}, y);$$

$$E_B: \qquad \text{take } \theta^{(m+1)} = g\left(\frac{1}{T}\sum_{j=1}^{T} x_{(m+1)j}, y\right).$$

We describe both of these algorithms as stochastic restoration–estimation (SRE) algorithms and denote them by SRE_A and SRE_B. There may be other variants of the SRE_B algorithm. When (5.1) defines the maximum (pseudo-) likelihood estimate, we refer to the algorithm as the SRM algorithm. Clearly, unless g is 'linear' the SRE_B algorithm, as stated, will be different from the SRE_A algorithm, which is more obviously motivated by (5.1), but this discrepancy often disappears asymptotically, as indicated later.

Example 5.1. Scalar exponential family

Suppose θ is scalar and

$$f(x, y \mid \theta) \propto (1/C(\theta)) \exp\{\theta H(x, y)\}.$$

Let

$$\theta = g(x, y) \tag{5.2}$$

denote the inverse of the likelihood equation, which is

$$C'(\theta)/C(\theta) = H(x, y).$$

Thus $g(x, y)$ is a function of the sufficient statistic, $H(x, y)$. With an abuse of notation, write (5.2) as

$$\theta = g(H(x, y)),$$

which motivates the following variant of the SRM_B algorithm,

$$\theta^{(m+1)} = g\left(\frac{1}{T}\sum_{j=1}^{T} H(x_{(m+1)j}, y)\right). \tag{5.3}$$

In general, g will often be a function only of the appropriate sufficient statistics, and certainly so if $\hat{\theta}$ in (5.1) is the maximum likelihood estimate.

Special cases, in the maximum likelihood context, are as follows.

(i) The case $T = 1$, for which E_A and E_B are equivalent, correspond to the SEM algorithm for mixture data described by Celeux & Diebolt (1985),

(ii) A further special case is discovered if $T \to \infty$ in example (5.1). In that case,

$$\frac{1}{T}\sum_{j=1}^{T} H(x_{(m+1)j}, y) \to E[H(X, y) \mid y, \theta^{(m)}],$$

so that the SRE_B algorithm corresponding to (5.3) is just the EM algorithm itself. For large T, the results should approximate those for the EM algorithm.

It is now appropriate to discuss the properties of the random sequence $\{\theta^{(m)}\}$. Let Ω_y denote the range of estimates $\theta^{(m)}$ for given y; Ω_y is assumed to be the same, for all m. Also, let

$$K(\cdot \mid \theta^{(m)}, y)$$

denote the conditional density of $\theta^{(m+1)}$, given $\theta^{(m)}$ and y. $K(\cdot\,|\,\theta, y)$ is then related to the conditional density of random variable $g(X, Y)$, given $Y = y$, with parameter θ, according to whichever of our algorithms is in use. Let

$$p_m(\cdot\,|\,y)$$

denote the conditional density of $\theta^{(m)}$ given $y, m = 1, 2, \ldots$. Then

$$p_{m+1}(\cdot\,|\,y) = \int_{\Omega_y} p_m(\tau\,|\,y)\,K(\cdot\,|\,\tau, y)\,\mathrm{d}\tau, \quad m = 1, 2, \ldots.$$

Often, $\{p_m\}$ will converge to the eigenfunction, corresponding to eigenvalue unity, of the integral operator

$$\int_{\Omega_y} K(\cdot\,|\,\tau, y)\,p(\tau\,|\,y)\,\mathrm{d}\tau,$$

in which $K(\lambda\,|\,\tau, y)$ can be regarded as the transition function of a homogeneous Markov chain.

By using the Hilbert projective metric, this convergence can be proved in the case where Ω_y is a closed, bounded region and $K(\lambda\,|\,\tau, y)$ is a positive-valued, continuous function of (λ, τ) on $\Omega_y \times \Omega_y$, provided the initial p_1 is continuous and positive-valued on Ω_y.

The required ergodicity of K cannot be established in any generality but the hope is that, in practice, $\{p_m\}$ converges to an eigendistribution p^* that satisfies

$$p^*(\cdot\,|\,y) = \int_{\Omega_y} K(\cdot\,|\,\tau, y)\,p^*(\tau\,|\,y)\,\mathrm{d}\tau \tag{5.4}$$

and that, 'eventually', $\theta^{(m)}$ can be regarded as being a random deviate from $p^*(\cdot\,|\,y)$. The resulting p^* can be used in many ways, of which the simplest is to use a sample mean of realizations as the 'final' estimate of θ. In general, it not possible to relate $p^*(\cdot\,|\,y)$ to the relationship between y and the true θ.

In this paper, we derive theoretical results only for a very simple example and merely illustrate the usefulness of the methods for Markov random fields.

Example 5.2

Suppose $X_i \sim N(\theta, \sigma_1^2)$, $Y_i\,|\,x_i \sim N(x_i, \sigma_2^2)$, $i = 1, \ldots, n$, that all $\{X_i\}$ are independent, and that (2.6) holds. We assume σ_1^2 and σ_2^2 are known, that the x_i are missing and that θ is of interest.

In this example, exact calculations are feasible, since, marginally, for each i,

$$Y_i \sim N(\theta, \sigma_1^2 + \sigma_2^2)$$

and

$$\hat{\theta} = \bar{y} = \frac{1}{n}\sum_{i=1}^{n} y_i$$

is the obvious estimate of θ from data $y_1, \ldots, y_n$.

SRM *algorithm with* $T = 1$

R-step: for $i = 1, \ldots n$, generate $x_i^{(m+1)}$ from $p(x_i\,|\,y_i, \theta^{(m)})$;

M-step: take $\theta^{(m+1)} = \dfrac{1}{n}\sum_{i=1}^{n} x_i^{(m+1)}$.

Table 1. *Values of $\bar{y}$, and sample mean and variance of late iterates of $\theta^{(m)}$, for two data-sets corresponding to Example 5.2*

data-set	$\bar{y}$	$\bar{\theta}^{(m)}$	$V(\theta^{(m)})$
1	1.4358	1.4266	0.0363
2	0.3315	0.3379	0.0292

Thus, in the general notation,

$$g(x, y) = \frac{1}{n} \sum_{i=1}^{n} x_i.$$

It is straightforward to show that, given y and $\theta^{(m)}$,

$$\theta^{(m+1)} \sim N\left[\frac{\sigma_2^2 \theta^{(m)} + \sigma_1^2 \bar{y}}{\sigma_1^2 + \sigma_2^2}, \frac{\sigma_1^2 \sigma_2^2}{n(\sigma_1^2 + \sigma_2^2)}\right]. \tag{5.5}$$

Relationship (5.5) defines the density $K(\cdot \mid \tau, y)$ and it is easy to check that the corresponding stationary density, $p^*(\cdot \mid y)$ satisfying (5.4) is that of the

$$N\left[\bar{y}, \frac{\sigma_2^2(\sigma_1^2 + \sigma_2^2)}{n(\sigma_1^2 + 2\sigma_2^2)}\right]$$

distribution. If, therefore, the SRM algorithm is carried out to convergence and a sample average of late iterates of $\theta^{(m)}$ is used as an estimate of the parameter, then that sample average should differ from the true maximum likelihood estimate only by an amount quantified by the above stationary distribution.

In a very simple illustration with $n = 20$, $\theta = 1$, and $\sigma_1^2 = \sigma_2^2 = 1$, two sets of data $\{y_i\}$ were generated. For each set, $\{\theta^{(m)}: m = 1, \ldots, 2000\}$ were generated and the sample mean and variance calculated for the last 1000 iterates are as given in table 1.

Note the proximity of $\bar{\theta}^{(m)}$ to $\bar{y}$ in each case, and that the theoretical variance is 0.0333.

SRM algorithm with general T

In this case, $\{x_{it}^{(m+1)}: i = 1, \ldots, n, t = 1, \ldots T\}$ are generated,

$$\theta^{(m+1)} = \frac{1}{nT} \sum_{i=1}^{n} \sum_{t=1}^{T} x_{it}^{(m+1)}.$$

Here $K(\cdot \mid \tau, y)$ is the density corresponding to a

$$N\left[\frac{\sigma_2^2 \tau + \sigma_1^2 \bar{y}}{\sigma_1^2 + \sigma_2^2}, \frac{\sigma_1^2 \sigma_2^2}{nT(\sigma_1^2 + \sigma_2^2)}\right]$$

random variable, and $p^*(\cdot \mid y)$ is that for a

$$N\left[\bar{y}, \frac{\sigma_2^2(\sigma_1^2 + \sigma_2^2)}{nT(\sigma_1^2 + 2\sigma_2^2)}\right]$$

random variable. As $T \to \infty$, $p^*(\cdot \mid y)$ becomes degenerate at $\bar{y}$, and the SRM algorithm becomes the EM algorithm.

We now present a case with a small sample size to show the difference that can arise between the $\mathrm{SRM_A}$ and $\mathrm{SRM_B}$ algorithms.

Example 5.3. Markov chain with periodic boundary condition and parameter β

Here
$$p(x_1, \ldots, x_5 \mid \beta) = \frac{1}{C(\beta)} \exp\left\{\beta \sum_{i=1}^{5} x_i x_{i+1}\right\},$$
where $x_i \in (-1, 1\}, i = 1, \ldots, 5, x_6 = x_1$, and $\beta > 0$. It is easy to show that
$$C(\beta) = 2(\mathrm{e}^{5\beta} + 10\mathrm{e}^{\beta} + 5\mathrm{e}^{-3\beta}).$$

Although we can clearly deal with the likelihood itself, consider estimating β by maximizing the pseudo-likelihood,
$$\prod_{i=1}^{5} p(x_i \mid x_{i-1}, x_{i+1}).$$

If
$$S_1(x) = \sum_{i=1}^{5} x_i x_{i+1}, \quad S_2(x) = \sum_{i=1}^{5} \delta(x_{i-1}, x_{i+1}),$$
it is easy to show that the maximum pseudo-likelihood estimate of β satisfies
$$\theta \equiv (\mathrm{e}^{2\beta} - \mathrm{e}^{-2\beta})/(\mathrm{e}^{2\beta} + \mathrm{e}^{-2\beta}) = S_1(x)/S_2(x).$$

Suppose now we observed, not x, but noisy data
$$y = (y_1, \ldots, y_5) = (0.80,\ 0.98,\ 1.01,\ -1.15,\ -0.95),$$
assumed to be modelled by $y_i = x_i + \epsilon_i$, where the $\epsilon_i \sim N(0, 0.36)$, independently.

In this very simple example, it is easy to generate samples from the posterior distribution, in order to implement the following iterative steps of the algorithms:

$\mathrm{SRM_A}$:
$$\theta^{(m+1)} = \frac{1}{T} \sum_{t=1}^{T} \frac{S_{1t}^{(m+1)}}{S_{2t}^{(m+1)}},$$

$\mathrm{SRM_B}$:
$$\theta^{(m+1)} = \sum_{t=1}^{T} S_{1t}^{(m+1)} \Big/ \sum_{t=1}^{T} S_{2t}^{(m+1)},$$

where $S_{1t}^{(m+1)}$ and $S_{2t}^{(m+1)}$ are S_1 and S_2 as obtained from the tth simulated x. The iterative step in the corresponding EM-algorithm is as follows,

EM:
$$\theta^{(m+1)} = \frac{E\{S_1(X) \mid y, \theta^{(m)}\}}{E\{S_2(X) \mid y, \theta^{(m)}\}}.$$

For the data provided as above, the EM algorithm (as given by the above formula) leads to the maximum (pseudo-)likelihood estimate $\hat{\theta}_{ML} = 0.9866$.

Both the $\mathrm{SRM_A}$ and $\mathrm{SRM_B}$ algorithms were run for 60 cycles, with $T = 1000$. The empirical means of the last 40 iterates were
$$\hat{\theta}_{\mathrm{A}} = 0.9954 \quad \text{and} \quad \hat{\theta}_{\mathrm{B}} = 0.9852,$$
in obvious notation. The latter is clearly close to $\hat{\theta}_{ML}$. Not surprisingly, $\hat{\theta}_{\mathrm{A}}$ is different from $\hat{\theta}_{\mathrm{B}}$, but is very close to the solution of
$$\theta^* = E\{S_1(X)/S_2(X) \mid \theta^*, y\},$$
namely, $\theta^* = 0.9955$.

Here, we did not estimate the variance of the noise. If the variance is larger, the difference between the estimates from the two procedures is larger.

Note that the sample size of $n = 5$ is very small. As $n \to \infty$, the ergodic properties of $S_1(X)$ and $S_2(X)$ will imply that $\hat{\theta}_{ML}$ and θ^* will ultimately be the same, so that the results of the SRM$_A$ and SRM$_B$ algorithms will be comparable.

Example 5.4. Binary first-order Markov chain

Suppose

$$f(x \mid \beta) = \frac{1}{C(\beta)} \exp\left\{\beta \sum_{i=1}^{n-1} \delta(x_i, x_{i+1})\right\},$$

where $x_i \in \{1, 2\}$ for all i. Given x_i, $Y_i \sim N(x_i, \sigma^2)$, independently for all i. Both β and σ^2 require to be estimated. For illustrative purposes, we consider maximum pseudo-likelihood estimation as the basis of the estimation of β. The log-pseudo-likelihood is

$$\beta S_1(x) - S_2(x) \ln(e^{2\beta}+1) - S_3(x)(\beta + \ln 2), \tag{5.6}$$

where
$$S_1(x) = \delta(x_1, x_2) + 2\sum_{i=2}^{n-2} \delta(x_i, x_{i+1}) + \delta(x_{n-1}, x_n),$$

$$S_2(x) = \sum_{i=2}^{n-1} \delta(x_{i-1}, x_{i+1}),$$

$$S_3(x) = n - 2 - S_2(x).$$

Maximization of (5.6) gives

$$\theta \equiv e^{2\beta}/(e^{2\beta}+1) = (S_1(x) - S_3(x))/2S_2(x). \tag{5.7}$$

Given x and y, σ^2 is estimated by

$$\sigma^2 = \frac{1}{n}\sum_{i=1}^{n} (y_i - x_i)^2. \tag{5.8}$$

For this example, the partition function can again be evaluated, and is

$$C(\beta) = 2(e^{\beta}+1)^{n-1} = 2[\surd((1-\theta)^{-1}-1)+1]^{n-1}.$$

Thus, the likelihood equation results in

$$e^{\beta}/(e^{\beta}+1) = S_4(x)/(n-1),$$

where
$$S_4(x) = \sum_{i=1}^{n-1} \delta(x_i, x_{i+1}).$$

In terms of θ, this gives

$$\theta = \frac{S_4^2(x)}{[n-1-S_4(x)]^2 + S_4^2(x)}. \tag{5.9}$$

Qian & Titterington (1990) developed, for this example, a recursive technique for maximizing the likelihood corresponding to the observed data, y. As a result, it is possible to compare the EM, SRM$_A$ and SRM$_B$ algorithm based on both the likelihood and the pseudo-likelihood functions. Results in Qian & Titterington (1990) showed

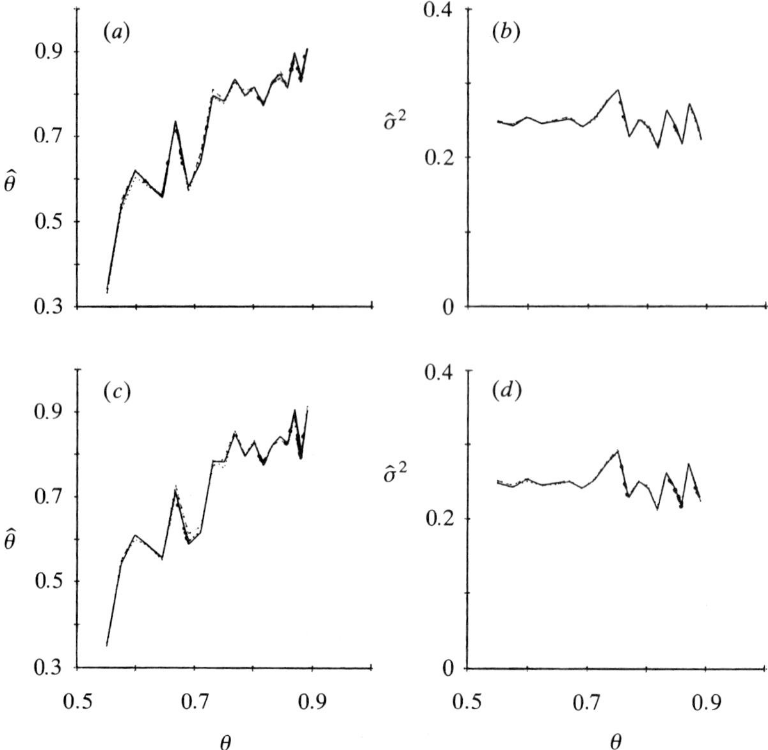

Figure 1. Comparison of EM, SRM_A and SRM_B algorithms for the estimation of θ and σ^2 in Example 5.4 with $\sigma^2 = 0.25$ and $n = 512$; (*a*), (*b*) based on likelihood; (*c*), (*d*) based on pseudo-likelihood. ——, EM; - - - - -, SRM_A; – – – –, SRM_B.

that the EM algorithm associated with the likelihood function provided very good parameter estimates. Here, we make comparisons with the other methods. (The comparison with the EM algorithm may offer useful extrapolation to the case of two-dimensional Markov random fields, for which it is impossible to carry out the EM algorithm.)

The ergodic properties of the statistics that appear in (5.7)–(5.9) should ensure that, for large n, the EM, SRM_A and SRM_B algorithms should perform similarly, whether the likelihood or the pseudo-likelihood is used. Accordingly, the value, 5, chosen for T was much less than was the case in Example 5.3.

Figure 1 displays results for $\sigma^2 = 0.25$ and $n = 512$. The value of $\theta = \text{e}^{2\beta}/(\text{e}^{2\beta}+1)$ was varied between 0.55 and 0.9. For each choice of parameters a single realization for x was generated and noise added to create y. It is clear that the results from the EM, SRM_A and SRM_B algorithms are very similar, and that the results based on likelihood are very similar to those based on pseudo-likelihood. For each procedure, except that associated the EM algorithm, 50 iterative cycles were carried out, and the quoted estimates are the averages of the last 40 cycles.

Figure 2 reports averages of the results based on 40 realizations of (x, y) for each choice of parameters, this time with $\sigma^2 = 0.09$ and $n = 64$. In this case, $T = 15$, 50 iterative cycles were carried out for each realization and the parameters were estimated by the averages of the final 30 iterates. Once again, all methods gave very similar results.

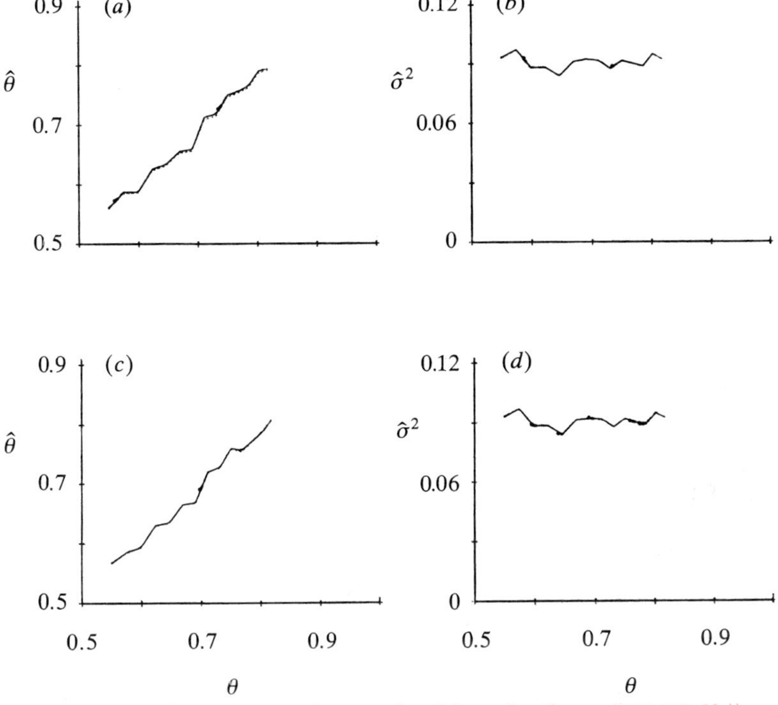

Figure 2. Comparison of EM, SRM$_A$ and SRM$_B$ algorithms for the estimation of θ and σ^2 in Example 5.4 with $\sigma^2 = 0.09$ and $n = 64$; (*a*), (*b*) based on likelihood; (*c*), (*d*) based on pseudo-likelihood ——, EM; , SRM$_A$; ----, SRM$_B$.

On a theoretical level, there are consistency results for maximum pseudo-likelihood estimation in the noise-free case (Geman & Graffigne 1987) but little has been established, as yet, for the case of noisy data.

6. An illustration using satellite data

The top four pictures in figure 3 display a four-band satellite image of the Lake of Menteith (Scotland's only 'Lake' rather then 'Loch'!) in Perthshire. As an illustrative exercise, we shall use our methodology to classify the pixels in 64×64 frame into a six-state scene. Thus, for each i, $x_i \in \{1, 2, \ldots, 6\}$. As a prior for X we use the first-order Markov random field, with probability function

$$p(x \mid \beta) = \frac{1}{C(\beta)} \exp\left\{\beta \sum_{i \sim j} \delta(x_i, x_j)\right\}, \tag{6.1}$$

where the summation is over nearest-neighbour pairs. It is arguable that this prior is unrealistically simple, in that it treats all pixel states symmetrically. The choice of six for the number of states is also somewhat arbitrary but serves well for illustrative purposes.

Each y_i is four-dimensional and it is assumed that, for a pixel in state $k (x_i = k)$,

$$y_i \sim N(\mu_k, V),$$

where V is a 4×4 covariance matrix. There are therefore 24 unknown parameters

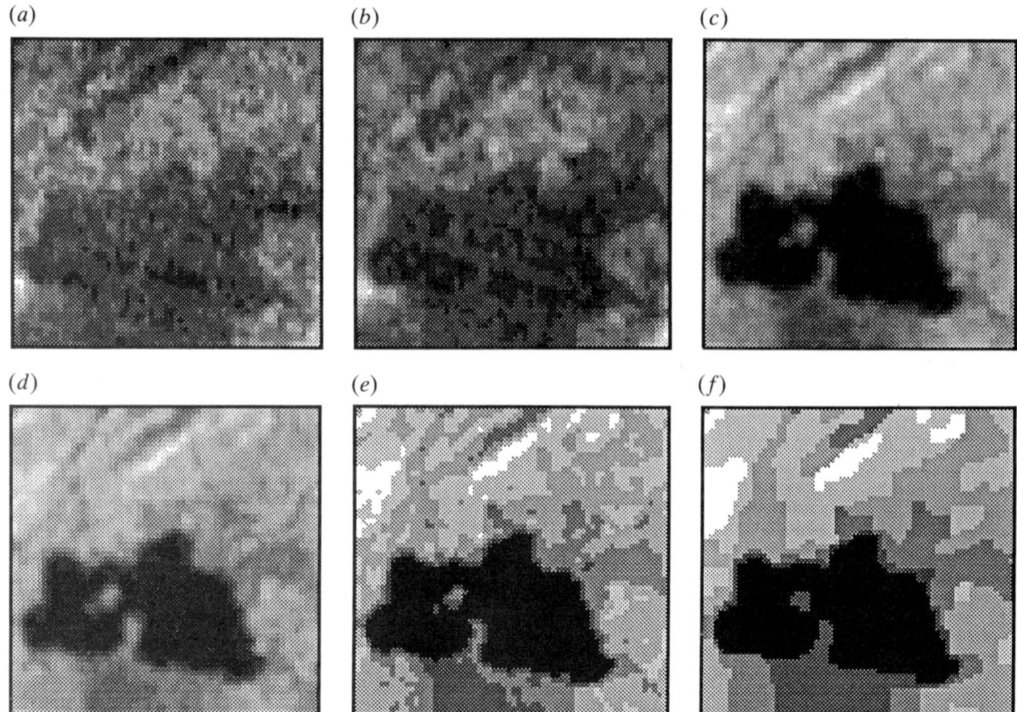

Figure 3. Satellite data (*a*)–(*d*) data from bands 1–4; (*e*) initial restoration from band 3 data; (*f*) final SRM restoration from band 3 data.

associated with the $\{\mu_k\}$ and 10 parameters including in V. Altogether, therefore, including the parameter β, there are 35 unknown parameters.

In the iterative procedure, we shall deal with β itself, rather than any transformation thereof, as we did for Examples 5.3 and 5.4.

The procedure is initiated by choosing a starting classification. In our illustration, we achieved this by taking the band 3 image alone, dividing the observed range into six equal ranges, and regarding these as the six states (a multi-state thresholding exercise.) The resulting restoration is presented in figure 3*e*. We report here only results from the SRM_A method, starting from the above restoration and using only the band-3 data. In this case there are only eight parameters (six means, one variance and β). The starting parameters were obtained by regarding the initial restoration as the true scene. The routine in each iterative stage was as follows.

(i) Run 50 cycles of the (posterior) Gibbs sampler starting from the current restoration.

(ii) Run the Gibbs sampler for further 100 cycles, estimating the parameters after each 10 cycles.

(iii) Obtain the average of the 10 sets of estimates and use them to start the next iteration. (Thus, in the previous notation, $T = 10$.)

Note that, at any stage, there is a current restored image. The restoration achieved after five of the above iterative cycles is displayed in figure 3*f*.

Next, we used this segmentation to initiate the SRM_A procedure using the four-dimensional data. Again, initial parameters were estimated by using this initial image. Note that the estimation of those mean-parameters and the covariance-

Table 2. *Satellite data: parameter estimates for mean vectors μ_k corresponding to six states $\{S1, \ldots, S6\}$, and for the covariance matrix V using (a) the* SRM$_A$ *algorithm, (b) the* PPL-ICM *algorithm, (c) the* PPL-EM-ICM *algorithm*

	(a) SRM parameters; $\hat{\beta} = 2.571$; $\{\hat{\mu}_k\}$					
	S1	S2	S3	S4	S5	S6
band 1	27.49	27.80	28.53	31.96	30.53	31.58
band 2	20.72	21.85	23.40	28.27	25.54	27.03
band 3	33.69	58.82	75.55	88.78	97.06	110.52
band 4	14.65	46.73	69.85	85.09	95.88	114.40
			$\hat{V}$			
		1.72	1.49	2.17	1.69	
		1.49	3.96	0.72	−0.75	
		2.17	0.72	27.42	45.81	
		1.69	−0.75	45.81	66.02	
	(b) PPL-ICM parameters; $\hat{\beta} = 2.736$; $\{\hat{\mu}_k\}$					
band 1	27.46	27.83	28.21	31.17	31.21	31.69
band 2	20.69	21.80	22.86	27.16	26.55	27.11
band 3	33.47	57.75	71.53	85.97	99.24	114.76
band 4	14.42	45.11	64.85	82.18	98.55	118.92
			$\hat{V}$			
		1.98	1.88	1.43	0.79	
		1.88	4.49	0.01	−1.55	
		1.43	0.01	24.72	32.10	
		0.79	−1.55	32.10	51.47	
	(c) PPL-EM-ICM parameters; $\hat{\beta} = 2.606$; $\{\hat{\mu}_k\}$					
band 1	27.47	27.82	28.23	31.21	31.18	31.65
band 2	20.70	21.80	22.89	27.23	26.48	27.03
band 3	33.54	57.91	71.80	86.23	99.30	114.61
band 4	14.50	45.33	65.16	82.46	98.65	118.84
			$\hat{V}$			
		1.98	1.88	1.62	1.02	
		1.88	4.47	0.34	−1.16	
		1.63	0.34	26.21	33.63	
		1.02	−1.16	33.63	52.98	

parameter is trivial. The algorithm was as above except that 30 iterative cycles were carried out, rather than 5. The averages of the parameter estimates obtained in the last 25 cycles are listed in table 2*a*.

It should be remarked that, in the SRM$_A$ algorithm as originally described, the T realizations of the posterior distribution are expected to be mutually independent. In the case of a two-dimensional Markov random field, it is computationally very expensive to generate many independent samples. However, two realizations, suitably far apart in a single operation of the Gibbs sampler, will be only loosely dependent, a phenomenon which we exploited in our calculations.

Once we have settled on a 'final' set of estimates of the parameters, in the present case the values in table 2*a*, there is more than one way of producing a 'final'

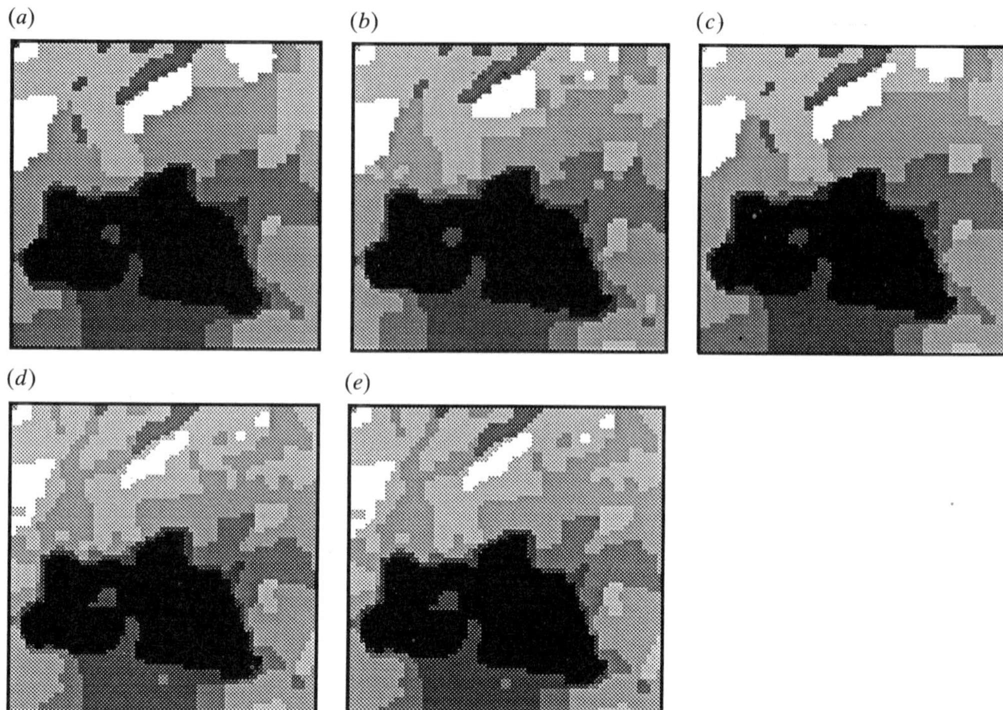

Figure 4. Satellite data: (*a*) final restoration from SRM phase; (*b*) ICM restoration using SRM estimates; (*c*) Gibbs sampler restoration using SRM estimates; (*d*) PPL-ICM restoration; (*e*) PPL-EM-ICM restoration.

classification for the pixels. Here, we report two methods, the former being the result of applying 25 cycles of Besag's (1986) ICM algorithm and the second involving implementation of an idea that originates in Besag *et al.* (1991). In the latter, many samples from the estimated posterior distribution are generated (we ran the Gibbs samplers for 500 cycles), and the most frequently generated state at any given pixel is used to create the restoration; in fact, we noted the states only after every fifth sampling cycle. We initiated both these procedures with the 'final' restoration created during the estimation phase. That particular image and the results of implementing the two restoration stages are presented in figure 4. The three images show very few mutual differences.

Also displayed in figure 4 are the results of two further estimation–restoration algorithms, the decision-directed PPL-ICM algorithm of Besag (1986) and the PPL-EM-ICM algorithm of Qian & Titterington (1991). Both were initiated from the initial restoration from the band 3 data, shown as figure 3*e*. The corresponding sets of parameter estimates resulting from the two procedures are shown in table 2*b*, *c*. The two sets of estimates are very similar, as are the two restorations. The reason for this may be that the noise is small compared with the 'original' image, bearing in mind the estimated covariance matrix and the mean-parameters; for example, for all three procedures, the differences of means between state 5 and state 6, associated with band 4 data, are about 20.0, while the standard deviations associated with band 4 are about 8.0 or less.

7. Discussion

In this paper, we have highlighted the difficulties involved in estimating parameters in hidden Markov models and we have reviewed methods that have been proposed for dealing with them. Detailed discussion and illustration have been presented of a subclass of the methods that have a strongly Monte Carlo flavour. Further research is required to consolidate these methods, in term of theory and implementation. However, the results reported here and, in particular, the account of Example 5.4, suggest that SRM algorithms based on pseudo-likelihood may be valuable alternatives to EM algorithms in problems in which the latter are totally impracticable.

The ideas should also be useful for the estimation of parameters in other contexts where models with interactive characteristics are essential. Prime examples are the fields of probabilistic expert systems (Lauritzen & Spiegelhalter 1988) and the use of statistical modelling in neural networks (see for instance Bourlard 1990).

W. Q. was supported by a research grant from the UK Science and Engineering Research Council, awarded under the Complex Stochastic System Initiative. The same agency funded the equipment on which the computational work was carried out. The authors are very grateful to Professor Julian Besag for his helpful comments on an earlier draft of the paper, and to Dr Jim Kay for kindly providing the data for the Lake of Menteith example.

References

Aykroyd, R. G. & Green, P. J. 1991 Global and local priors, and the location of lesions using gamma-camera imagery. *Phil. Trans. R. Soc. Lond.* A **337**, 323–342. (This volume.)

Bartlett, M. S. 1971 Physical nearest-neighbour models and non-linear time-series. *J. appl. Prob.* **8**, 222–232.

Besag, J. E. 1975 Statistical analysis of non-lattice data. *Statistician* **24**, 179–195.

Besag, J. E. 1976 Parameter estimation for Markov fields. Tech. Rep. 198, Series 2. Dept Statistics, Princeton Univ.

Besag, J. E. 1986 On the statistical analysis of dirty pictures (with discussion). *Jl R. statist. Soc.* B **48**, 259–302.

Besag, J. E., York, J. & Mollie, A. 1991 Bayesian image restoration, with two applications in spatial statistics (with discussion). *Ann. Inst. Statist. Math.* **43**, 1–59.

Bourlard, H. E. 1990 How connectionist models could improve Markov models for speech recognition. In *Advanced neural computers* (ed. R. Eckmiller), pp. 247–254. Amsterdam: North Holland.

Celeux, G. & Diebolt, J. 1985 The SEM algorithm: a probabilistic teacher algorithm derived from the EM algorithm for the mixture problem. *Comput. Statist. Q.* **2**, 73–82.

Chalmond, B. 1989 An iterative Gibbsian technique for reconstruction of M-ary images. *Pattern Recognition* **22**, 747–761.

Dempster, A. P., Laird, N. M. & Rubin, D. B. 1977 Maximum likelihood estimation from incomplete data via the EM algorithm (with discussion). *Jl R. statist. Soc.* B **39**, 1–38.

Derin, H. & Elliott, H. 1987 Modelling and segmentation of noisy and textured images using Gibbs random fields. *IEEE Trans. Pattern Analysis Machine Intell.* **PAMI-9**, 39–55.

Frigessi, A. & Piccioni, M. 1988 Parameter estimation for two-dimensional Ising fields corrupted by noise. Quaderno no. 12, IAC-CNR, Rome.

Frigessi, A. & Piccioni, M. 1990 Consistent parameter estimation for 2-D Ising fields corrupted by noise: numerical experiments. In *Proc. IMS Conf. Spatial Statistics and Imaging* (ed. A. Possolo). Haywood, California: Institute of Mathematical Statistics.

Geman, S. & Geman, D. 1984 Statistical relaxation, Gibbs distributions and the Bayesian restoration of images. *IEEE Trans. Pattern Analysis Machine Intell.* **PAMI-6**, 721–741.

Geman, S. & Graffigne, C. 1987 Markov random field image models and their applications to computer vision. In *Proc. Int. Congress of Mathematicians* (ed. A. M. Gleason), pp. 1498–1517. Berkeley, California: American Mathematical Soc.

Geman, S. & McClure, D. 1987 Statistical methods for tomographic image reconstruction. *Bull. Int. statist. Inst.* (BK. 4) **52**, 5–21.

Geyer, C. J. & Thompson, E. A. 1992 Constrained Monte Carlo maximum likelihood for dependent data (with discussion). *Jl R. statist. Soc.* B **54**. (In the press.)

Gray, A. J., Kay, J. W. & Titterington, D. M. 1991 On the estimation of noisy binary Markov random fields. (Submitted.)

Greig, D. M., Porteous, B. T. & Seheult, A. H. 1989 Exact maximum a posterior estimation for binary images. *Jl R. statist. Soc.* B **51**, 271–279.

Juang, B. H. & Rabiner, L. R. 1991 Hidden marker models for speech recognition. *Technometrics* **33**, 251–272.

Lauritzen, S. L. & Spiegelhalter, D. J. 1988 Local computations with probabilities on graphical structures and their application to expert systems (with discussion). *Jl R. statist. Soc.* B **50**, 157–224.

Little, R. J. A. & Rubin, D. B. 1983 On jointly estimating parameters and missing values by maximizing the complete data likelihood. *Am. Statist.* **37**, 218–220.

Little, R. J. A. & Rubin, D. B. 1987 *Statistical analysis with missing data.* New York: Wiley.

Marriott, F. H. C. 1975 Separating mixtures of normal distributions. *Biometrics* **31**, 767–769.

Moyeed, R. A. & Baddeley, A. J. 1991 Stochastic approximation of the MLE for a spatial point pattern. *Scand. J. Statist.* **18**, 39–50.

Pickard, D. K. 1987 Inference for discrete Markov fields: the simplest nontrivial case. *J. Am. statist. Ass.* **82**, 90–96.

Possolo, A. 1986 Estimation of binary Markov random fields. Tech. Rep. no. 77, Dept Statistics, University Washington, Seattle.

Ripley, B. D. 1988 *Statistical inference for spatial processes.* Cambridge University Press.

Qian, W. & Titterington, D. M. 1990 Parameter estimation for hidden Gibbs chains. *Statist. Prob. Lett.* **10**, 49–58.

Qian, W. & Titterington, D. M. 1991 Stochastic relaxations and EM algorithms for Markov random fields. *J. statist. Comput. Simul.* **41**. (In the press.)

Smith, A. F. M. 1991 Bayesian computational methods. *Phil. Trans. R. Soc. Lond.* A **337**, 369–386. (This volume.)

Titterington, D. M. 1984 Comments on a paper by S. C. Sclove. *IEEE Trans. Pattern Analysis Machine. Intell.* **PAMI-6**, 656–658.

Titterington, D. M. 1990 Some recent research in the analysis of mixture distributions. *Statistics* **21**, 619–641.

Titterington, D. M., Smith, A. F. M. & Makov, U. E. 1985 *Statistical analysis of finite mixture distributions.* London and New York: Wiley.

Veijanen, A. 1990 An estimator for imperfectly observed Markov fields. Research Rep. no. 74, Dept Statistics, University of Helsinki.

Younes, L. 1988 Estimation and annealing for Gibbsian fields. *Ann. Inst. H. Poincare* **24**, 269–294.

Younes, L. 1989 Parametric inference for imperfectly observed Gibbsian fields. *Prob. Theory Rel. Fields* **82**, 625–645.

TURBULENT FLOW STRUCTURE NEAR WALLS

Compiled and edited by J.D.A. Walker

Over the past decade, considerable progress has been made in determining the nature of turbulent flow near walls. These advances have occurred through innovative new experimental methodologies, direct numerical simulations, and significant new theoretical developments. Many of these aspects are described in this volume.

Incompressible channel flow and the constant pressure turbulent boundary layer represent canonical bounded turbulent flows and here the emerging dynamical picture is that of a flow dominated by vortex motion and its effects (Grass *et al.*; Falco; Smith *et al.*). Note that a recent synopsis of the direct numerical simulations of turbulence is given elsewhere (S.K. Robinson, *A. Rev. Fluid Mech.* **23**, 601–639 (1991)). An objective assessment of direct numerical simulations is given by Zang, while Perry *et al.* describe a closure model based on flow structure concepts. As yet, the study of flow structure in non-canonical situations is in its infancy. However, there is reason to believe that similar physical phenomena occur (in modified form) in more complex environments. The situations discussed here are: (1) high-speed compressible boundary layers (Smits); (2) separated flows (Simpson); and (3) drag reducing flows (Harder & Tiederman).

175 pages Paperback ISBN 0 85403 442 0

First published in *Philosophical Transactions of the Royal Society,* Series A, Vol. 336, 1991

Price including packing and postage
£19.50 (UK addresses) £21.00 (Overseas addresses)

The Royal Society, 6 Carlton House Terrace, London SW1Y 5AG

THE PHYSICS OF SOLAR FLARES

Organized and edited by J.L. Culhane and C. Jordan

During the period of the 1980 solar maximum three space missions (*P78-1*, *Solar Maximum Mission* and *Hinotori*) carrried out extensive studies of solar flares. In their different ways all of these missions contributed significant new information to our understanding of the solar flare phenomenon.

In this volume the contribution made by these three spacecraft to the study of the energy release and the related creation of high-temperature plasma, the transport of energy from the primary release site, the production of rays at energies up to 10 MeV and the ejection of solar matter into interplanetary space are reviewed.

Discussions of the current theoretical basis of magnetic energy conversion, the role of magnetic loops in solar flares and the acceleration of electrons and protons in the implusive phase are presented. In addition the relation between ground-based optical observations of solar flares and aspects of the X-ray data is assessed. All three spacecraft included high-resolution X-ray crystal spectrometers. The use of these instruments in establishing the properties of the high-temperature plasma is described and the status of the atomic data required for the interpretation of the spectra is evaluated.

For the 1991 solar maximum the Japanese *Solar-A* satellite will carry four major instruments that will allow observations to be undertaken with a unique combination of high spatial and spectral resolution over a wide range of X-ray and γ-ray energies. The nature of this mission is described and its role in advancing our understanding of the solar flare problem discussed.

176 pages paperback ISBN 0 85403 444 7

First published in Philosophical Transactions of the Royal Society,
Series A, Vol. 336, 1991

Price including packing and postage
£19.50 (UK addresses) £21.00 (Overseas addresses)

The Royal Society
6 Carlton House Terrace
London SW1Y 5AG